Bibliographies on the History of Science and Technology
Vol. 20

THE HISTORY OF AGRICULTURAL SCIENCE AND TECHNOLOGY

Garland Reference Library of the Humanities
(Vol. 1371)

Bibliographies on the History of Science and Technology

ROBERT MULTHAUF
Smithsonian Institution
Series Editor

THE HISTORY OF AGRICULTURAL SCIENCE AND TECHNOLOGY

An International Annotated Bibliography

R. Douglas Hurt
Mary Ellen Hurt

GARLAND PUBLISHING, INC.
New York & London / 1994

Library of Congress Cataloging-in-Publication Data

Hurt, R. Douglas
The history of agricultural science and technology : an international annotated bibliography / by R. Douglas Hurt and Mary Ellen Hurt.
p. cm. — (Bibliographies on the history of science and technology ; v. 20) (Garland reference library of the humanities ; vol. 1371)
Includes index.
ISBN 0-8240-7182-4
1. Agriculture—History—Bibliography. 2. Agriculture—Bibliography. I. Hurt, Mary Ellen. II. Title. III. Series. IV. Series: Garland reference library of the humanities ; vol. 1371.
Z5071.H87 1994
[S419]
016.63'09—dc20 93-28006
CIP

Printed on acid-free, 250-year-life paper
Manufactured in the United States of America

FOR

ADLAI AND AUSTIN

GENERAL INTRODUCTION

This bibliography is one of a series designed to guide the reader into the history of science and technology. Anyone interested in any of the components of this vast subject area is part of our intended audience, not only the student, but also the scientist interested in the history of his own field (or faced with the necessity of writing an "historical introduction") and the historian, amateur or professional. The latter will not find the bibliographies "exhaustive," although in some fields he may find them the only existing bibliographies. He will in any case not find one of those endless lists in which the important is lumped with the trivial, but rather a "critical" bibliography, largely annotated, and indexed to lead the reader quickly to the most important (or only existing) literature.

Inasmuch as everyone treasures bibliographies it is surprising how few there are in this field. Justly treasured are George Sarton's *Guide to the History of Science* (Waltham, Mass., 1952; 316 pp.), Eugene S. Ferguson's *Bibliography of the History of Technology* (Cambridge, Mass., 1968; 347 pp.), François Russo's *Histoire des Sciences et des Techniques, Bibliographie* (Paris, 2nd ed., 1969; 214 pp.), and Magda Witrow's *ISIS Cumulative Bibliography. A bibliography of the history of science* (London, 1971–; 2131 pp. as of 1976). But all are limited, even the latter, by the virtual impossibility of doing justice to any particular field in a bibliography of limited size and almost unlimited subject matter.

For various reasons, mostly bad, the average scholar prefers adding to the literature, rather than sorting it out. The editors are indebted to the scholars represented in this series for their willingness to expend the time and effort required to pursue the latter objective. Our aim has been to establish a general framework which will give some uniformity to the series, but otherwise to leave the format and contents to the author/compiler. We have urged that introductions be used for essays on "the state of the field," and that selectivity be exercised to limit the length of each volume to the economically practical.

Since the historical literature ranges from very large (e.g., medicine) to very small (chemical technology), some bibliographies will be limited to the most important writings while others will include modest "contributions" and even primary sources. The problem is to give useful guidance into a particular field—or subfield—and its solution is largely left to the author/compiler.

In general, topical volumes (e.g., chemistry) will deal with the subject since about 1700, leaving earlier literature to the area of chronological volumes (e.g., medieval science); but here, too, the volumes will vary according to the judgment of the author. The topics are international, with a few exceptions, but the literature covered depends, of course, on the linguistic equipment of the author and his access to "exotic" literatures.

Robert Multhauf
Smithsonian Institution
Washington, D.C.

CONTENTS

PREFACE

This bibliography is designed for advanced students and scholars who are working in a variety of topical areas in the history of agricultural science and technology. Specifically, this bibliography emphasizes the agricultural science and technology related to grain, livestock, forage and fiber production. Because of space limitations, we have not been able to address the historical literature on horticulture, viticulture, apiculture or other subject areas such as sugar cane, vegetables and nuts. We have, however, included important chapters on the history of veterinary medicine, entomology, nutrition, women and the Green Revolution in addition to more traditional subjects. By so doing, we have provided an introduction to the literature for historical research in agricultural science and technology. Moreover, our approach is international. We have sought to include a selection of the most important sources published during the last twenty years as well as the major studies published earlier. No bibliography can be complete, of course, that attempts to survey the historical literature that will be useful for historians, scientists, geographers, sociologists and political scientists on an international basis. We believed, however that these 1380 citations for 48 topical areas will provide the researcher with a place to start.

Our bibliography takes a different approach than most studies of this nature. Rather than provide annotated citations merely to books and articles that primarily have been written by historians, we have surveyed the scientific and social science literature that will be useful for historical research. Moreover, most of the articles that we have cited have not been published in historical journals. Consequently, this bibliography will provide an introduction to non-traditional, historical publications that historians might otherwise overlook. Simply put, we believe that while scholars who are not historians take a different methodological and analytical approach to their research and writing, their work, whether specifically historical or not, can be exceptionally useful for historians and other scholars engaged in research and writing about the history of agricultural science and technology.

We are also convinced that anyone who conducts research in the fields of agricultural science and technology must be knowledgeable about accessing the recent literature in the most efficient way. In this respect, we have placed considerable emphasis on database sources. These electronic sources are essential for any serious researcher to help identify the most recent studies that are published in a multiplicity of journals, bulletins, working papers and books. Contact with an agricultural information specialist, familiar with the **Directory of On-Line Databases**, is important when designing a literature search strategy. In addition, as we prepared this bibliography, two critical titles, **Current Bibliography in the History of Technology**, and **Current Bibliography of the History of**

Science, became available in electronic formats. Researchers should check with their library's history reference specialist for access to these titles, which are available through the Research Libraries Information Network (RLIN) in the United States.

Although this bibliography has been topically arranged to include the most important subjects related to the history of agricultural science and technology, space prevented the duplication of titles that legitimately could be located in several sections. Consequently, the researcher should consult the index and remain flexible in mind. Some citations, for example, are located under the heading of genetics, but they easily could also have been placed in the chapter on biotechnology. Similarly, historical sources on hog cholera are located in the chapter on veterinary medicine rather than in the section on swine. We have placed each citation where we believe it should most logically be found. Usually, the placement of a citation depended on our design, but occasionally we made the determination arbitrarily because of necessity. As a result, the researcher should consult both the general as well as subject specific categories. Subject specific bibliographies, for example, will be found in the specific chapters and sections, while more general bibliographies will be located in the chapter devoted to those sources.

We have made a concerted effort to exclude purely economic literature, although economics influences the development of agricultural science and technology. Moreover, we have included many general sources that are not specifically related to agricultural science. We believe, however, that these sources are essential for the researcher. We have, for example, provided citations for general works in botany, genetics, and microbiology, among others, that will help provide the necessary background for anyone conducting research in many areas of agricultural science. In addition, we have included chapters on institutions and policy, both of which play a fundamental role in the development of agricultural science and technology and thereby the history of each. Most of our citations, however, involve technical histories. At the same time, we have emphasized published sources that the researcher should be able to acquire with relative ease. While most our our sources have been published in English, we have included the most recent and important publications in other languages, especially French, German, Spanish, Polish and Russian.

Last, with the dissolution of the former Soviet Union, a number of information vendors have recently contracted to provide bibliographic access to the agricultural science and technology literature in the Russian language. The researcher should check the current edition of the **Directory of On-Line Databases** for these citations. This procedure will be an important first step in any literature search for foreign language publications.

ACKNOWLEDGEMENTS

Anyone who conducts research and writing for an annotated bibliography understands the often tedious and time-consuming nature of this work. Fortunately, several individuals, Stephanie Carpenter, Peter Limas and Sheri Ruehling, who served as our research assistants, made our task considerably easier. We are thankful for their help at the computer and in the library. Without that assistance, we would not have been able to undertake, let alone complete, this bibliography. In addition, Gerry McKiernan provided essential reference help at the William Robert and Ellen Sorge Parks Library at Iowa State University, and the Interlibrary Loan Department maintained its reputation for considerable skill and efficiency.

The History of Agricultural Science and Technology

CHAPTER I

INDEXING AND ABSTRACTING SERVICES

1. **Abstracts on Tropical Agriculture**.
   Amsterdam, The Netherlands: Royal Tropical Institute (KIT), Information and Documentation (ID), 1975-

   A monthly abstracting service providing access to world-wide literature concerned with sub-tropical and tropical agriculture. Provides good coverage of African countries. Formerly titled **Tropical Abstracts.** Includes geographic, plant taxonomic and author indexes. Abstracts are in English. Available on-line as **TROP** and on CD-ROM as **TROPAG & RURAL**.

2. **Agricultural Engineering Index**.
   St. Joseph, MI: American Society of Agricultural Engineers (ASAE), 1907-

   A keyword index to **Agricultural Engineering** as well as to a few selected foreign agricultural engineering titles. Indexing also is included for the society's transactions and technical papers. The citations are arranged by subject, then by year of publication. Volumes are available for 1907-1960, 1961-1970, 1971-1980, 1981-1985 and 1986-1990. These cumulations are kept up-to-date by the annual indexing found in the **Annual Comprehensive Index of Publications**, also published by the ASAE.

3. **Agrindex**. Rome: Food and Agriculture Organization (FAO), International Information System for the Agricultural Sciences and Technology (AGRIS), Coordinating Centre, 1975-

   A monthly index to world agricultural literature, including the FAO publications. Valuable for information on developing countries and nations whose agricultural literature is not covered well by the **Bibliography of Agriculture** or the numerous **CAB Abstracts**. With support from the United Nations, as well as other international organizations, citations are collected from more than 90 countries by numerous regional centers throughout the world. Citations are to "conventional library materials" as well as patents, maps, charts, films and computer media. Topically arranged with author and subject indexes. Available on CD-ROM as well as on-line as **AGRIS**.

4. **Agroselekt**. Berlin, Germany: Akademie-Verlag, 1955-

   A monthly international agricultural abstracting service issued in four parts: **Reihe 1: Landtechnik** (agricultural engineering); **Reihe 2: Pflanzenproduktion** (crops, agronomy); **Reihe 3: Tierproduktion** (animal husbandry); and **Reihe 4: Veterinaermedizin** (veterinary medicine). Formerly **Landwirtschaftliches Zentralblatt** (1955-1984) also issued in four series.

5. **America: History and Life**. Santa Barbara: ABC-CLIO, 1964-

   A quarterly international abstracting service that accesses articles, books and dissertations written on American and Canadian history. It was issued in four parts from 1974 until 1988. Available on CD-ROM as well as on-line.

6. **Applied Science and Technology Index**. New York: H.W. Wilson, 1913-

   A monthly indexing service that provides access to articles published in the English language. Primarily emphasizes American engineering and applied sciences. Formerly **Industrial Arts Index**, 1913-1957. Available on CD-ROM and on-line as well as magnetic tape.

7. **Bibliographia Historiae Rerum Rusticarum Internationalis.** Budapest: Museum Rerum Rusticarum Hungariae Budapestini, 1961-

   A biennial classified bibliography of articles, books, papers and dissertations in agricultural history published in the "major languages of the world." Geographic index provided.

8. **Bibliography of Agriculture**. Phoenix: Oryx Press, 1942-

   A monthly index providing international coverage for all aspects of agricultural research. Detailed review of American agricultural research provided through access to experiment station publications and other esoteric materials deposited with the National Agricultural Library (NAL) in Beltsville, Maryland. Earlier indexing of the American experiment station reports is found in the **Experiment Station Record** published between 1889 and 1946. Available on CD-ROM, on-line or magnetic tape as **AGRICOLA**. Access to earlier American agricultural research station reports is found by checking the **Experiment Station Record** (1889-1946).

9. **Bibliography of the History of Medicine**. Bethesda, MD: National Library of Medicine (NLM), 1964-

   Annual index to the history of medicine including veterinary medicine. In three parts: Part I, Biographies; Part II, Subject Index; and Part III, Authors. Available on CD-ROM and on-line as well as magnetic tape as **HISTLINE**.

10. **Biological Abstracts**. Philadelphia: BioSciences Information Service, 1926-

    A semi-monthly international abstracting service for the life sciences. This is a two-part service with the periodical literature in the above title while the proceedings are abstracted in **Biological Abstracts/RRM**. Extensive indexing is provided. Available on CD-ROM, on-line and magnetic tape as **BIOSIS**.

11. **Biological and Agricultural Index: A Cumulative Subject Index to Periodicals in the Field of Biology and Agriculture and Related Sciences**. New York: H.W. Wilson, 1916-

A monthly indexing service that provides coverage of the basic biological and agricultural journals, published in the English language. Formerly **Agricultural Index** 1916-1964. Available on CD-ROM, on-line or magnetic tape.

12. **CAB Abstracts**. Wallingford, Oxon, United Kingdom: CAB International, 1973-

An extensive, international abstracting service, updated monthly, that classifies the literature from more than 8,500 journals covering the numerous sub-fields of agricultural sciences, including veterinary science. The beginning date for each abstracting title varies. Available on CD-ROM, on-line and magnetic tape as **CAB Abstracts**. The following classified list provides an example of the titles in this vast series.

ANIMAL PRODUCTION

Animal Breeding Abstracts
Dairy Science Abstracts
Nutrition Abstracts and Reviews, Series B: Livestock Feeds and Feeding
Pig News and Information
Poultry Abstracts

ENGINEERING

AgBiotech News and Information
Agricultural Engineering Abstracts
Irrigation and Drainage Abstracts

PLANT PROTECTION

Biocontrol News and Information
Chickpeas and Pigeonpeas
Cotton and Tropical Fibres
Crop Physiology Abstracts
Faba Bean Abstracts
Field Crop Abstracts
Horticultural Abstracts
Lentil Abstracts
Maize Abstracts
Nematological Abstracts
Phosphorus in Agriculture
Plant Growth Regulator Abstracts
Potato Abstracts
Review of Agricultural Entomology
Review of Plant Pathology
Seed Abstracts
Soils and Fertilizers
Sorghum and Millet Abstracts
Soyabean Abstracts
Tropical Oil Seeds
Wheat, Barley, and Triticale Abstracts
Weed Abstracts

VETERINARY MEDICINE

Animal Disease Occurrence
Helminthological Abstracts
Index Veterinarius
Protozoological Abstracts
Review of Medical and Veterinary Entomology
Review of Medical and Veterinary Mycology
Veterinary Bulletin

13. **Chemical Abstracts.** Columbus: American Chemical Society, 1907-

    A world-renowned international chemistry database, updated weekly and available on-line since 1967. Extensive indexing includes formula, ring systems and patents. Available on-line as **CA Search**.

14. **Current Contents**. Philadelphia: Institute for Scientific Information (ISI), 1969-

    A weekly current awareness service that is based on the table of contents of journals in science and technology. Two series of this service are of particular interest to agricultural historians, **Current Contents: Agriculture, Biology, and Environmental Sciences**, and **Current Contents: Engineering, Technology and Applied Sciences**. This information is then combined with the citations found in the **Science Citation Index** to form the database **SciSearch** which is available on CD-ROM, on-line and magnetic tape.

15. **Engineering Index**. Hoboken, NJ: Engineering Information Inc., 1884-

    A monthly, international abstracting service essential to accessing the engineering literature. Available on CD-ROM, on-line and magnetic tape as **Compendex**. Access to publications may also be found in the **Classed Subject Catalog** of the Engineering Societies Library. This catalog was published in 1963.

16. **FAO Documentation: Current Bibliography**. Rome: Food and Agriculture Organization (FAO), 1967-

A monthly index to FAO documents, with annual cumulated author and subject indexes. The documents are available, in microfiche as **FAO Documentation**. In addition, there are cumulated indexes in microfiche for 1976-1979 and 1980-1990. Earlier printed indexes are **FAO Documentation: Current Index** (1967-1971); **FAO Documentation; Quarterly List of Publications and Main Documents** (1961-1967); and **United Nation's Document Index** (1950-1962). The FAO documents also have been indexed by **Agrindex** since 1975, and citations can be searched on-line with **AGRIS**.

17. **Francis Bulletin Signaletique 522, Histoire des Sciences et des Techniques**. Nancy, France: Centre National de la Recherche Scientifique (CNRS), Institut de l'Information Scientifique et Techniques (INIST), 1940-

A quarterly international index with some abstracts. Provides access to books and articles covering the history of science and technology including agricultural history. From 1961-1990 this source was titled **Bulletin Signaletique 522**. Prior to 1961, it was called **Bulletin Analytique**. Available on CD-ROM, on-line and magnetic tape as **FRANCIS: Histoire es Sciences et des Techniques**.

18. **Historical Abstracts**. Santa Barbara: ABC-CLIO, 1955-

A quarterly international abstracting service for historical literature excluding American and Canadian history. It is published in two parts: **Part A, Modern History (1450-1914)** and **Part B, Twentieth-Century (1914-)**. Available on CD-ROM and on-line as well as magnetic tape.

19. **Index Medicus**. Bethesda, MD: National Library of Medicine (NLM), 1879-

    A monthly international index to the medical literature, including veterinary medicine. Since 1961, it has been cumulated annually by the **Cumulative Index Medicus**. Between 1927-1956, it was cumulated by the **Quarterly Cumulative Index Medicus**. Available on CD-ROM, on-line and magnetic tape as **MEDLINE**.

20. **International Bibliography of Historical Sciences**. Edited for the International Committee of Historical Sciences. Munich: K.G. Saur, 1926-

    Biennial international indexing service for the history of the sciences including agricultural science and technology. Includes a detailed geographic index. Issues are alternately published in the following languages: French; German; Italian; and Spanish.

21. **PASCAL**. Nancy, France: Centre National de la Recherche Scientifique (CNRS), Institut de l'Information Scientifique et Techniques (INIST), 1940-

A monthly international abstracting service that provides access to books and articles covering the literature in agriculture, science, technology and engineering. Issued in more than 30 series, with the following two series being important to historians of agricultural science and technology: **PASCAL THEMA 280: Sciences Agronomiques, Productions Vegetales**; and **PASCAL THEMA 215: Biotechnologies**. This abstracting service was titled **Bulletin Signaletique** (1955-1984); and **Bulletin Analytique** (1940-1955). Available on CD-ROM, on-line and magnetic tape as **PASCAL**.

22. **Referativnyi Zhurnal**. Moscow: All Union Institute of Scientific and Technical Information (VINITI), 1954-

A monthly international abstracting service issued in 64 sections providing access to the literature in the sciences and technology, including agricultural science and technology. This title is available on-line as **VINITI** and will soon be accessible outside the Commonwealth of Independent States. The following sections produced by the Vesesoyuznyi Institut Nauchnoy i Teknicheskoy Informatsii (VINITI) in Moscow may be useful to agricultural historians:

Pochvovedenie I Agrokhimiya (Soil Science and Agricultural Chemistry)
Rastenievodstvo (Plant Breeding)
Traktory I Sel'skokhozyaistvennye Masiny I Orudiya (Tractors and Farm Machinery and Equipment)

Zhivotnovodstvo I Veterinariya (Animal Husbandry and Veterinary Science)

23. **Science Citation Index**. Philadelphia: Institute for Scientific Information (ISI), 1955-

    A monthly index to the scientific literature divided into three sections: a keyword index; a source index; and a citation index. The citation index enables a researcher to locate articles that have been cited for particular scholars. Five-year cumulations have been published. Available on CD-ROM, on-line and magnetic tape as **SciSearch**.

CHAPTER II

BIBLIOGRAPHIES AND CATALOGS

24. Alvear, Alfredo. **Bibliografia de Bibliografias Agricolas de America Latina**. 2a ed rev. y ampl. Turrialba, Costa Rica: Instituto Interamericano de Ciencias Agricolas (IICA), Biblioteca y Servicio de Documentacion, 1969. 121 pp.

    A classified, annotated bibliography on agricultural topics including veterinary medicine. Includes author index.

25. Bebee, Charles N. **Bibliography of American Agricultural Bibliographies, 1984: A Categorized Listing of Bibliographies Indexed in AGRICOLA**. Bibliographies and Literature of Agriculture (BLA), No. 54. Beltsville, MD: National Agricultural Library, 1986. 159 pp.

    This is the latest of four bibliographies of bibliographies compiled by the author from the AGRICOLA database between 1977 and 1984 and published as part of the Bibliographies and Literature of Agriculture series.

The first bibliography (1977) was published as BLA-1, the second (1978-1982) as BLA-32, and the third (1983) as BLA-44. An author index is included.

26. Besterman, Theodore. **Agriculture: A Bibliography of Bibliographies.** Totowa, NJ: Rowman and Littlefield, 1971. 302 pp.

    Compiled by the publisher from the 4th edition of Besterman's 5 volume set, **A World Bibliography of Bibliographies**, which covered bibliographies issued up to 1963. Although dated, this classified bibliography is international in scope, then arranged geographically by country when appropriate. Citations are in the language in which the work was published and there are no annotations.

27. **Bibliography of Agricultural Bibliographies, 1961-70, 1971-72**. Zug, Switzerland: Inter-Documentation Co., 1973, 1977. 163 microfiche in 3 boxes, 59 microfiche in 1 box.

    Contains bibliographic entries for items received by the library of the Landbouwhogeschool Wageningen (Netherlands). During this same time frame the **Pudoc Bulletin** (1960-1984) was published, which provided access to the Dutch agricultural literature.

28. Black, George W. **American Science and Technology: A Bicentennial Bibliography**. Carbondale, IL: Southern Illinois University Press, 1979. 170 pp.

A compilation of over one thousand citations to works published during the American bicentennial year, 1976, by historians as well as scientists.

29. Bowers, Douglas. **A List of References for the History of Agriculture in the United States, 1790-1840**. Davis, CA: Agricultural History Center, University of California-Davis, 1969. 141 pp.

    This bibliography is part of the series of bibliographies produced by the Agricultural History Center which builds on the work of Everett E. Edwards (1930). It contains a section on "improving agriculture" which provides citations to articles and books dealing with agricultural science, technology, and education.

30. Bush, Ernest Alfred Radford. **Agriculture: A Bibliographical Guide**. 2 Vols. London, MacDonald and Jane's, 1974.

    An extensive bibliography containing over 9000 annotated entries. The author intended to update the work of Blanchard (1958) and Lauche (1957), so the coverage is primarily new titles published between 1958-1971. Arranged by topics with author index, and detailed subject index.

31. Carlson, L., ed. "Bibliography of the History of Australian Science," in **Historical Records of Australian Science**. 1981-

    This bibliography has appeared annually starting with Volume 5 no. 2 (November,

1981). Some entries are annotated and provide access to books, journal articles, and conference papers. Citations are for the natural as well as the applied sciences, including agriculture. Additional items may be found in the **Records of the Australian Academy of Science** (1966-1980).

32. Chen, Qiubo, and Charles Bebee. **Chinese Publications in the Collections of the National Agricultural Library: A Bibliography.** 2 Vols. Bibliographies and Literature of Agriculture series, No. 80. Beltsville, MD: National Agricultural Library, 1989.

    The author, a visiting scholar from China, researched and compiled this bibliography during his visit to the National Agricultural Library. Volume 1 contains classified citations to 3000 items on Chinese agriculture, written in either Chinese or English, and available through the **AGRICOLA** database. Volume 2 contains an additional 1500 citations later added to the **AGRICOLA** database. Additional serial titles relating to Chinese agriculture may be located in the **Dictionary Catalog of the NAL, 1862-1965**.

33. "Current Bibliography in the History of Technology," in **Technology and Culture**. One issue annually. 1964-

    An annual bibliography issued by the Society for the History of Technology published in their journal, **Technology and Culture**. The bibliography is international in scope and includes author and subject indexes. Usually published in the July issue, however, as of 1992, this bibliography will be published as a separate issue. In addition, the bibliography has expanded to cover citations of dissertations as well as conference papers. Also the bibliography is now available electronically through the Research Libraries Information Network (RLIN).

34. Dean, Genevieve Catherine. **Science and Technology in the Development of Modern China: An Annotated Bibliography**. London: Mansell Information Publishing, 1974. 265 pp.

    A survey of western literature on the history of science and technology in China. A good place to start.

35. **Dictionary Catalog of the National Agricultural Library.** 73 Vols. Totowa, NJ: Rowman and Littlefield, 1967-70.

    Provides author, title, subject access to the collections. Volume 73 contains an author index to translations. A supplement, for items added between 1966-1984, was published as the **National Agricultural Library Catalog**, Vol. 1-19. For access to more recent materials use the **Bibliography of Agriculture** or the on-line database **AGRICOLA**.

36. Edwards, Everett Eugene. "A Bibliography of the History of Agriculture in the United States." United States Department of Agriculture, **Miscellaneous Publications No. 84**. 1930. 307 pp.

This classic bibliography originated in 1927, when it was first developed as a guide to the literature for a USDA graduate-level class on the history of agriculture. Emphasis is on materials published between 1900 and 1929.

37. _____. "References on Agricultural History as a Field for Research." United States Department of Agriculutre, **Bibliographic Contributions No. 32**, 1937. 41 pp.

Historians began to view American agricultural history as a field of study as early as 1904. Includes a chronological list of many of the important first articles in this field of study. Descriptive annotations are provided.

38. _____. "Selected References on the History of English Agriculture." United States Department of Agriculture, **Bibliographical Contributions, No. 24**. 1935. 42 pp.

An early bibliography of British agricultural history. Everett recommends titles for beginning the study of the history of English agriculture. Includes limited citations for science and technology.

39. Fusonie, Alan M. **Heritage of American Agriculture: A Bibliography of pre-1860 Imprints**. NAL Library List No. 98. Beltsville, MD: National Agricultural Library, 1975. 71 pp.

A selective bibliography of books and periodicals. Includes items published by early agricultural societies. Arranged in three parts: books by authors; agricultural society transactions by region and state; and a list of early periodical titles held by the NAL.

40. Fussell, George Edwin. **Agricultural History in Great Britain and Western Europe Before 1914: A Discursive Bibliography**. London: Pindar Press, 1983. 157 pp.

    A narrative rather than a bibliographic listing in which Fussell reflects on the major books dealing with European agricultural history. Written with his customary wit. Emphasizes England, Germany, Belgium, Luxembourg, Italy, France and the Netherlands. Essential for an overview of European agricultural historiography.

41. Hall, Carl W. **Bibliography of Agricultural Engineering Books.** ASAE Publication, No. 8-76. St. Joseph, MI: American Society of Agricultural Engineers, 1976. 84 pp.

    A useful, brief, international bibliography, arranged by author. Includes subject index.

42. _____. **Bibliography of Bibliographies of Agricultural Engineering and Related Subjects**. ASAE Publication, No. 9-76. St. Joseph, MI: American Society of Agricultural Engineers, 1976. 62 pp.

A classified, international bibliography to supplement the above bibliography.

43. **Historia Agriculturae: Tentative Bibliography**. Groningen: Nederlands Agronomisch-Historisch Instituut, 1971-1975. 471 pp.

    A short-lived, international bibliography to agricultural history literature published in European languages in the early 1970's, and includes a geographic index. This bibliography was published earlier as **Historia Agriculturae, Yearbook** (1953-1976). This same institute published the periodical **Historia Agriculturae** (1953-1978).

44. **ISIS Cumulative Bibliography: A Bibliography of the History of Science**. London: Mansell, History of Science Society, 1971-

    Formed from the "ISIS Critical Bibliography", the collective set of cumulative bibliographies provides access to much to the international literature on the history of science from 1913-1985. Issued in the following sets: 1-90, (covering 1913-65) was published in 1971 in 6 volumes; 91-100, (covering 1965-1975) was published in 1980 in 2 volumes; and 101-110 (covering 1975-1985) was published by G.K. Hall (Boston, MA) in 1989 in 2 volumes.

45. "ISIS Current Bibliography," in **ISIS: an International Review Devoted to the History of Science and its Cultural Influences**. One issue annually, 1989-

Formerly "ISIS Critical Bibliography", 1913-1988. Annual bibliographies providing access to the international literature in the history of science have been published by the History of Science Society with their journal **ISIS** since 1913. The annual bibliographies were considered "critical" while George Sarton, the founder of **ISIS** was the compiler. After his retirement in 1953, however, the annual bibliographies became increasingly selective rather than critical. As a result, in 1989, the Society changed the title of its annual bibliography to "ISIS Current Bibliography." Since 1992, this bibliography has been available in electronic format from Research Libraries Information Network (RLIN).

46. Lauche, Rudolf. **Internationales Handbuch der Bibliographien des Landbaues**. Munich, Bayerischer Landwirtschaftsverlag, 1957. 411 pp.

A classic agricultural bibliography especially for European agricultural science and technology.

47. Lawani, S.M., F.M. Alluri, and E.N. Adimorah. **Farming Systems in Africa: A Working Bibliography**. Boston: G.K. Hall, 1979. 251 pp.

Includes nearly 2,000 titles, many dealing with agricultural science and technology. Topically arranged.

48. Moore, Julie L. **The Updata Index to U.S. Department of Agriculture, Agricultural Handbooks, Numbers 1-540**. Los Angeles, CA: Updata Publications, 1982. 80 pp.

An index to the handbooks published by the USDA from 1949-1979. Provides title and subject index, as well as document number index. The index uses common names rather than Latin. The handbooks themselves cover a number of topics, from pesticides to anatomy, that are often found in government documents collections under the SUDOC number A1.76:+.

49. Morris, V.J., and D.J. Orton. "List of Books and Pamphlets on Agrarian History" in **Agricultural History Review**. One issue annually, 1953-

An annual bibliography has been published in the British Agricultural History Society's journal since its inception, with the first list prepared by George E. Fussell. Various historians have prepared the list over the years, and it usually appears in Part II. The list covers books and pamphlets from the previous year. A bibliographical essay entitled "Annual List and Brief Review of Articles on Agrarian History" analyzes the periodical literature, and it is usually published in Part I.

50. Naftalin, Mortimer L. **Historic Books and Manuscripts Concerning General Agriculture in the Collection of the National Agricultural Library**. NAL Library List, No. 26. Washington, D.C.:

USDA, National Agricultural Library, 1967. 94 pp.

Brief annotations to pre-1800 European items as well as pre-1830 American titles.

51. **Publications on International Agricultural Research and Development**. Washington, D.C.: The Consultive Group on International Agricultural Research (CGIAR), 1983. 386 pp.

CGIAR has joined with the International Rice Institute (IRRI), a CGIAR supported center, and the German Agency for Technical Cooperation (GTZ) to issue an annual catalog to promote international agricultural research publications. This citation is for the first catalog only as place of publication will change. For the historian, this provides a quick overview of the available publications from the various international agricultural research agencies.

52. **Pure and Applied Science Books, 1876-1982**. 6 Vols. New York: Bowker, 1982.

An excellent starting place for the agricultural historian, with 220,000 entries to books, including foreign titles available in the United States. Science and engineering are covered, with a section on agriculture and veterinary science as well. This is a classified list, with entries in the Library of Congress cataloging style. Volume 6 provides an extensive author as well as title index.

53. Pursell, Carroll W., and Earl M. Rogers. **A Preliminary List of References for the History of Agricultural Science and Technology in the United States**. Davis, CA: Agricultural History Center, University of California-Davis, 1966. 46 pp.

An expansion of the work done by Everett E. Edwards in 1930. However, this bibliography deals with agricultural science and technology, especially the production side. The science portion of this bibliography was later updated by Rossiter (1980) while the technology portion was updated by Whitehead (1979).

54. Richardson, R. Alan, and Bertrum H. MacDonald. **Science and Technology in Canadian History: A Bibliography of Primary Sources to 1914.** Research Tools for the History of Canadian Science and Technology, No.3. Thornhill, Canada: HSTC Publications, 1987. 105 microfiches.

Covers more than 58,000 entries to articles and monographs in science, engineering, and technology, published by Canadians or about Canada. Includes information on agriculture and veterinary medicine. Location codes are included, with most items available from libraries in Ottawa and Quebec.

55. Rogers, Earl M., and Rogers, Susan H. "Significant Books on Agricultural History Published" in **Agricultural History**. One issue per year, 1977-

This bibliography has appeared in the Agricultural History Society's [U.S.A.] publication for many years, usually in the fall issue. Prior to 1977, the bibliography was entitled "Books on Agricultural History."

56. Rossiter, Margaret W. **A List of References for the History of Agricultural Science in America**. Davis, CA: Agricultural History Center, University of California-Davis, 1980. 62 pp.

    Building on the work of Pursell and Rogers (1966) this bibliography deals specifically with agricultural science. Sections cover chemistry and plant and animal sciences as well as institutions.

57. Rothenberg, Marc. **The History of Science and Technology in the United States: A Critical and Selective Bibliography**. New York: Garland Publishing, 1982. 242 pp.

    An annotated bibliography which includes titles on agricultural science and technology as well as many other citations. Includes a section on women.

58. Rumney, Thomas A. **The Geography of World Agriculture: A Selected Bibliography**. 3 Vols. Public Administration Series: Bibliography, P-2949-2951. Monticello, IL: Vance Bibliographies, 1990.

    Covers the period 1945-1990. Volume 1 provides general works, while volume 2 emphasizes the Western Hemisphere and volume 3 the Eastern Hemisphere. Arranged by region.

59. Russo, Francois. **Elements de Bibliographie de l'Histoire des Sciences et des Techniques**. 2nd ed. Paris, Hermann, 1969. 214 pp.

A revised and updated version of his earlier work, "Histoire des Sciences et des Techniques: Bibliographie," **Actualites Scientifiques et Industrielles** 1204 (1954). This international bibliography includes citations on agricultural mechanization and plant sciences.

60. Schlebecker, John T. **Bibliography of Books and Pamphlets on the History of Agriculture in the United States, 1607-1967**. Santa Barbara, CA: ABC Clio, 1969. 183 pp.

An introduction to American agricultural history including agricultural science and technology.

61. Smit, Pieter. **History of the Life Sciences: An Annotated Bibliography**. Amsterdam: Asher, 1974. 1071 pp.

Over 4000 well-annotated entries are provided on the history of the life sciences. Most of the entries are for books published prior to 1971, and the scope is international.

62. Spellman, Noel, and Katharine Mochon. **Agricultural History: An Updated Author, Title and Subject Index to the Journal, 1977-1990**. Davis, CA: Agricultural History Center, University of California-Davis, 1991. 49 pp.

An essential index to **Agricultural History**, the journal of record for the field in the United States.

63. Tucher, Andrea J. **Agriculture in America, 1622-1860: Printed Works in the Collections of the American Philosophical Society, the Historical Society of Pennsylvania, the Library Company of Philadelphia.** America to 1860 series, Vol. 2. New York: Garland, Publishing, 1984. 212 pp.

    A catalog of printed works related to agriculture, published prior to 1861 and held by the three research libraries in Philadelphia. Excellent subject index, as well as chronology.

64. Weintraub, Irwin. **Black Agriculturalists in the United States 1865-1973, An Annotated Bibliography**. Bibliographical Series, No. 7. University Park, PA: The Pennsylvania State University Library, 1976. 317 pp.

    An annotated bibliography emphasizing the contributions made by black agriculturists. Author and subject indexes provided.

65. White, K.D. **A Bibliography of Roman Agriculture**. Bibliographies in Agricultural History, No. 1. Reading, England: University of Reading, Institute of Agricultural History, 1970. 63 pp.

    A classified, annotated bibliography for works located in the United Kingdom, Rome and Paris.

66. Whitehead, Vivian B. **A List of References for the History of Agricultural Technology**. Davis, CA: Agricultural History Center, University of California-Davis, 1979. 76 pp.

Building on the work of Pursell and Rogers (1966), the bibliography deals specifically with agricultural technology. Covers sources of power, fertilizers, and equipment.

67. _____. **Agricultural History: An Index, 1927-1976**. Davis, CA: Agricultural History Center, University of California-Davis, 1977. 100 pp.

An index to the first fifty volumes of **Agricultural History**. In three sections: table of contents; authors; and subjects.

68. Zahlan, Antoine Benjamin. **Agricultural Bibliography of Jordan, 1974-1983: Selected, Classified & Annotated**. London: Ithaca Press, 1984. 80 pp.

Contains 291 annotated references that primarily concern economic and social development. However, see this bibliography for mechanization, irrigation, pest and plant protection, veterinary medicine and animal husbandry.

69. _____. **Agricultural Bibliography of Sudan, 1974-1983**. London: Ithaca Press, 1984. 325 pp.

A selected and annotated bibliography based on research by the author as well as the **CAB Abstracts** and **AGRIS** databases. Detailed coverage with sections on agricultural mechanization and specific crops.

CHAPTER III

REFERENCE WORKS

DICTIONARIES

70. **Academic Press Dictionary of Science and Technology**. New York: Academic Press, 1992. 2,240 pp.

    See for agricultural, chemical, ecological, veterinary and biotechnological terminology. Covers 124 fields of science and technology.

71. **Agricultural Terms: English-Chinese (Nung yeh t'zu hui: Ying Han tui chao)**. Terminology Bulletin No. 40/C. Rome: Food and Agriculture Organization of the United Nations, 1977. 313 pp.

    A glossary providing translations from English to Chinese characters. Access to other terminology bulletins published by the Food and Agricultural Organization (FAO) is provided by **Agrindex** or **AGRIS** on-line.

72. **Chihabi's Dictionary of Agricultural and Allied Terminology: English-Arabic, with an Arabic English Glossary**. Beirut: Librarie du Liban, 1982. 907 pp.

A dictionary based on the work of Emir Moustapha El Chihabi in his **Dictionnaire Francais-Arabe des Termes Agricoles**. In alphabetical order by terms, both common and Latin.

73. Dalal-Clayton, D. Barry. **Black's Agricultural Dictionary**. 2nd ed. London: A. & C. Black, 1985. 432 pp.

Defines 3,500 terms used in British agriculture by farmers and researchers. Includes illustrations.

74. Fisher, John Lionel. **A Medieval Farming Glossary of Latin and English Words, Taken Mainly From Essex Records.** London: Published for the Standing Conference for Local History by the National Council of Social Service, 1968. 41 pp.

A brief but useful guide to the Latin terminology used in British farming between 1200 and 1600. Based on manorial records from the Essex area.

75. Grant, Roger, and Clair Grant, ed. **Grant and Hackh's Chemical Dictionary Containing the Words Generally Used in Chemistry, and Many of the Terms Used in Related Sciences**. 5th ed. New York: McGraw-Hill, 1987. 641 pp.

Valuable for chemical terminology used in agriculture as well as engineering. Includes formulas.

76. Haensch, Gunther, and Haberkamp de Anton Gisela. **Dictionary of Agriculture: German, English, French, Spanish, Italian, and Russian, Systematical and Alphabetical**. [**Worterbuch der Landwirtschaft**]. 5th ed. Amsterdam: Elsevier Scientific Pub. Co., 1986. 1,264 pp.

    A glossary, now in six languages, arranged systemically, rather than alphabetically. Separate language indexes are provided to assist users. This is a basic working title for agricultural historians involved with international research.

77. King, R. C., and W.D. Stansfield. **A Dictionary of Genetics**. 4th ed. New York: Oxford University Press, 1990. 406 pp.

    Useful for scholars writing on the newly emerging areas of plant and animal genetics. Includes a chronology of the major developments in genetics.

78. MacKay, Susan E. **Field Glossary of Agricultural Terms in French and English, Emphasis: West Africa.** West Lafayette, IN: International Programs in Agriculture (IPIA) and Department of Languages and Literature, Purdue University, 1984. 103 pp.

    This "field" glossary, in two parts, French-English and English-French, provides access to Africa-specific agricultural terms. Similar glossaries are often published by agricultural research centers or by governmental agencies including the Peace Corps.

79. **McGraw-Hill Dictionary of Scientific and Technical Terms.** 4th ed. New York: McGraw-Hill, 1989. 2,088 pp.

    The most comprehensive and easy to use dictionary providing brief definitions for terms in 100 fields of science and technology, including agriculture. Common names as well as Latin terms are used. Formulas are given when appropriate.

80. Schlebecker, John T. **The Many Names of Country People: An Historical Dictionary From the Twelfth Century Onward.** New York: Greenwood Press, 1989. 325 pp.

    A useful dictionary for the varying terminology that farmers have used in daily life, including words relating to agricultural science and technology.

81. Vashkhnil, I., V.G. Kozlovsky, and N.G. Rakipov. **English-Russian Dictionary of Agriculture** [**Anglo-Russkii Selskokhoziaistvennyi Slovar**]. Oxford: Pergamon, 1985. 875 pp.

    A standard glossary for translating British and American agricultural terminology to Russian. Alphabetical order using common as well as Latin terms.

82. Winburne, John N., ed. **A Dictionary of Agricultural and Allied Terminology.** East Lansing, MI: Michigan State University Press, 1962. 905 pp.

    The standard American agricultural dictionary which should be in the personal library of all agricultural historians. Emphasis on common names rather than Latin terms. Appendix A includes a brief bibliography by subject.

GUIDES TO THE LITERATURE

83. Blanchard, J. Richard, and Louis Farrell, eds. **Guide to Sources for Agricultural and Biological Research.** Berkeley: University of California Press, 1981. 735 pp.

    An excellent source for students and researchers interested in agricultural history. International in scope, it provides detailed annotations to bibliographies as well as reference works. An entire section is devoted to agricultural history. This is a revised and updated version of **Literature of Agricultural Research** by Blanchard and Ostvold (1958).

84. Cloud, Gayla Staples. **Selective Guide to Literature on Agricultural Engineering.** Engineering Literature Guides, No. 4. Washington, D.C.: American Society for Engineering Education, Engineering Libraries Division, 1985. 23 pp.

A short, but useful, annotated guide providing a starting point for research in the history of agricultural engineering.

85. Ferguson, Eugene S. **Bibliography of the History of Technology.** Cambridge: Society for the History of Technology, 1968. 347 pp.

    More than a bibliography, this work is a guide to the print and non-print literature available to researchers. The second half of the book provides an annotated bibliography.

86. Hall, Carl W., and Wallace C. Olsen. **The Literature of Agricultural Engineering**. Ithaca: Cornell University Press, 1992. 432 pp.

    Analyzes the trends in agricultural engineering during the past century with emphasis since 1950. The book is part of the Core Agricultural Literature Project and, as of August, 1992, an agreement had been reached between the Rockefeller Foundation and University Microfilms, Inc. to produce a compact disk library which will include the full-text of monographs and journals. An essential guide. All research in this field should begin here.

87. Hurt, Charlie Deuel. **Information Sources in Science and Technology**. Englewood, CO: Libraries Unlimited, 1988. 362 pp.

    An easy to use guide to more than 2,000 titles, primarily reference books. Arranged by discipline and by type of reference book. Entries are annotated. Includes author/title index.

88. Jayawardene, S. A. **Reference Books for the Historian of Science: A Handlist.** Occasional Publications, No. 2. London: Science Museum, 1982. 229 pp.

    Useful handbook that leads researchers to international sources. Includes author, title and subject indexes.

89. Lambert, Jill. **How to Find Information in Science and Technology**. 2nd ed. London: Library Association, 1991. 108 pp.

    A recent publication that provides information on both British as well as American titles. Also covers titles in compact disk and on-line formats. Includes title/subject index.

90. Levick, George R. T., ed. **Guide to Agricultural Information Sources in Asia and Oceania.** FID Publication, No. 592. The Hague: Federation Internationale de Documentation (FID) 1980. 72 pp.

    Although dated this title provides a starting point. Arranged geographically with international sources first. The bibliography is organized by type of information and by subject. The descriptions include scope, output, and publisher information. More up-to-date information is available from the International Service for National Agricultural Service (ISNAR) in The Hague.

91. Lilley, George P. **Information Sources in Agricultural and Food Science.** London: Butterworths, 1981. 603 pp.

An excellent handbook for researchers in agricultural history, includes a chapter by J.S. Creasey on "Agrarian and Food History." International in scope. Includes annotated entries and index.

92. Morgan, Bryan. **Keyguide to Information Sources in Agricultural Engineering.** London: Mansell Pub., 1985. 209 pp.

An excellent guide to the literature that includes bibliographical essays as well as an annotated bibliography. Provides names and addresses for contacts around the world. This is a starting place for scholars of the history of agricultural engineering.

93. Mount, Ellis, and Beatrice Kovacs. **Using Science and Technology Information Sources.** Phoenix, Oryx Press, 1991. 189 pp.

Designed as a companion guide to guides to the literature. Mount provides direction for understanding the types of literature and its use for research in science and technology. Mount has published a great deal about the literature of science and technology. He is the author of the **Guide to Basic Information Sources in Engineering** (1976).

94. Owen, Dolores B. **Abstracts and Indexes in Science and Technology.** 2nd ed. Metuchen, N.J.: Scarecrow Press, 1985. 235 pp.

    Provides a detailed description of abstracts and indexes in print as well as electronic formats. Excellent, detailed coverage of various **CAB** databases. Arranged by subject area and alphabetically. Includes title index.

95. Sarton, George. **Horus: A Guide to the History of Science; A First Guide for the Study of the History of Science, with Introductory Essays on Science and Tradition.** Waltham, MA: Chronica Botanica Co., 1952. 316 pp.

    A classic work by the founder of **ISIS**. This first guide serves many functions including providing an index to earlier **ISIS** volumes. Later editions of the guide have been published as supplements to **ISIS**. This guide is an extension of a bibliography found in his **Study of the History of Science** (1936).

96. Szilard, Paula. **Food and Nutrition Information Guide.** Littleton, CO: Libraries Unlimited, 1987. 358 pp.

    An excellent source for researchers interested in the history of human nutrition, food science and technology. Annotated entries with author, title and subject indexes.

97. Wolford, Albert John. **Science and Technology.** Vol. 1 of **Guide to Reference Materials.** London: The Library Association, 1989. 791 pp.

    An indispensable guide to the international literature in the area of agricultural science and technology. Provides annotated entries to books as well as articles and chapters of books.

BIOGRAPHICAL GUIDES

98. **Agricultural and Veterinary Sciences International Who's Who.** 3rd ed. 2 Vols. Harlow, Great Britain: Longman Group U.K., 1987.

    Previously published as **Who's Who in World Agriculture,** this edition provides detailed biographical sketches of 7,500 "senior" agricultural scientists and veterinarians. Coverage is international with emphasis on British scientists. Country and subject indexes are provided. Also available from Gale Research Co. in Detroit, Michigan, as part of the Reference on Research Series.

99. **American Men and Women of Science, 1992-1993**. 8 Vols. 18th ed. New Providence, NJ: Bowker, 1992.

    Contains brief biographical sketches on current agricultural scientists arranged in alphabetical order by name. Date and place of birth, major field of study, place and date of education, honorary degrees, memberships and research addresses are given when possible. Volume 8 contains an extensive index which is organized by discipline, state and name.

100. **Biographical Encyclopedia of Scientists**. 2 Vols. New York: Facts on File, 1981.

These volumes contain more than 2,000 entries on scientists with descriptions of their lives and work. A useful chronology in volume 2 from 590 B.C. to 1981 is classified by discipline. Agricultural scientists are found within biology and chemistry.

101. Cannon, Grant G. **Great Men of Modern Agriculture**. New York: Macmillan, 1963. 256 pp.

Illustrated biographical essays about nineteen men from around the world, who made considerable contributions to the development of modern agriculture. Includes a glossary of agricultural terms, bibliography and index.

102. **Dictionary of Scientific Biography**. 16 Vols. New York, Charles Scribner's Sons, 1981.

Short essays on international agricultural researchers who are deceased. Essays are signed. Volume 16 contains an index and scientists are listed by discipline.

103. Dies, Edward Jerome. **Titans of the Soil: Great Builders of Agriculture**. Chapel Hill: University of North Carolina Press, 1949. 213 pp.

Provides biographical information on seventeen agriculturists who made monumental contributions to agriculture. Included are John Deere and Edmund Ruffin.

104. Elliott, Clark A. **Biographical Dictionary to American Science: The Seventeenth Through the Nineteenth Centuries**. Westport, CT: Greenwood Press, 1979. 360 pp.

Essentially includes scientists born between 1606 and 1867, the latter of whom had died by the early twentieth century. Includes bibliographical information.

105. Fruton, Joseph S. **A Bio-Bibliography for the History of the Biochemical Sciences Since 1800**. Philadelphia: American Philosophical Society, 1982. 885 pp.

An excellent source for citations to biographical information on biologists and agricultural scientists. Arranged alphabetically by name. Citations are for both monographs and serials. A supplement was published in 1985.

106. Ivins, Lester S., and A.E. Winship. **Fifty Famous Farmers**. New York: Macmillan, 1924. 407 pp.

Written for the layman. Discusses farmers as inventors and breeders of improved plants and animals. Includes soil scientists.

107. Kruif, Paul de. **Hunger Fighters**. New York: Harcourt, Brace and Company, 1928. 377 pp.

A popular work on the contributions of the agricultural scientists Mark Carleton, Marion Dorset, John Mohler, George Harrison Shull, Stephen Babcock, and Joseph Goldberger.

108. Moore, Ernest G. "Men Who Went Before." In **Science in Farming: Yearbook of Agriculture 1943-1947**, 1-16. Washington, D.C.: United States Department of Agriculture, 1947.

A brief look at men who made contributions in "farm science." Additional information is found in the 1962 yearbook **Men and Milestones**, which also includes information on a woman, Louise Stanley.

109. **National Academy of Sciences: Biographical Memoirs**. Washington, D.C.: National Academy Press, 1877-

Includes essays on selected scientists who were members of the Academy. Selected bibliographies. Cumulative index issued in vol. 61 (1992).

110. Rasmussen, Wayne D. "USDA Scientists: Selected Profiles, Challenges and Achievements." **Journal of NAL Associates** 8 no. 1-4 (1983): 24-34.

Despite recent criticism of the narrowness and inefficiency of their research programs, scientists in the United States Department of Agriculture have contributed much to the improvement of agricultural productivity.

111. Williams, Trevor I. **Biographical Dictionary of Scientists**. 3rd ed. London: A.C. Black, 1982. 672 pp.

Brief biographical sketches of deceased scientists, written and signed by scholars. Appendix includes additional scientists cited in sketches but for whom no full biographies are included. Also contains a chronology for birth and death dates from 624 B.C. to A.D. 1981. Subject index encompasses agricultural as well as veterinary medicine.

GENERAL REFERENCE

112. **Abbreviations Used by FAO for International Organizations, Congresses, Commissions, Committees, etc.** Terminology Bulletin No. 27/rev. 4. Rome: Food and Agriculture Organization of the United Nations, 1988. 205 pp.

A useful guide in its fourth revision since 1974. Provides an explanation of acronyms used by the FAO. Although arranged in alphabetical order in English, it provides the French and Spanish abbreviations as well as translations. Indexed in all three languages. Available from UNIPUB, Lanham, Maryland.

113. **Agricultural Research Centres: A World Directory of Organizations and Programs**. 10th ed. Essex, Great Britain: Longman Group U.K., 1990. 987 pp.

A directory of private, governmental, and academic research centers that conduct or fund research in the many sub-fields of agriculture and veterinary medicine. Indexed by organization and by research topic. Also available from Gale Research Co. in Detroit, Michigan.

114. **Agricultural Statistics** Washington, D.C.: Government Printing Office, 1936-

A standard handbook that provides statistics for all facets of American agriculture. An essential title for every scholar of agricultural history in the United States.

115. **Directory of On-Line Data Bases**. Detroit, MI.: Cuadra/Gale Research, 1979-

A serial publication of critical importance to serious researchers in agricultural history. Used by professional on-line searchers to keep aware of the vast amount of information available in on-line databases. Similar directories are also available for "portable" databases, i.e. compact disks.

116. **Directory of Technical and Scientific Directories: A World Bibliographic Guide to Medical, Agricultural, Industrial, and Natural Science Directories**. 6th ed. Harlow, England: Longman Group U.K., 1989. 302 pp.

Provides titles, descriptions and subscription information. Arranged by sections, such as agriculture, environment, engineering and medical and biological sciences. Within sections, the arrangement is geographical with the U.S. under North America.

117. **FAO Production Yearbook**. Rome: Food and Agriculture Organization of the United Nations, 1947-

Provides international agricultural statistics.

118. Golob, Richard, and Eric Brus. **The Almanac of Science and Technology: What's New and What's Known**. Boston: Harcourt Brace Jovanovich Publishers, 1990. 530 pp.

This title includes an article on "Genetic Engineering in Agriculture," which is useful for the layperson and covers the literature until 1988.

119. Hellemans, Alexander, and Bryan H. Bunch. **Timetables of Science: A Chronology of the Most Important People and Events in the History of Science.** New York: Simon and Schuster, 1988. 656 pp.

A useful reference for timelines of major developments. Includes explanatory essays.

120. Heynen, William J. **Agricultural Maps in the National Archives of the United States, 1860-1930.** National Archives and Records Service, Reference Information Leaflet, No. 75. Washington, D.C.: Government Printing Office, 1976. 25 pp.

A research source that historians of American agricultural science and technology should not overlook in order to gain another perspective on their work.

121. Lapedes, Daniel N. **McGraw-Hill Encyclopedia of Food, Agriculture and Nutrition.** New York: McGraw-Hill, 1977. 732 pp.

An important reference for the scholar, student and layperson about all aspects of agriculture, food manufacturing and nutrition. Includes technological as well as economic and political analysis. See for various topics, such as irrigation, pesticides, cultivation, harvesting and the Green Revolution.

122. **McGraw-Hill Encyclopedia of Science and Technology**. 7th ed. 20 Vols. New York: McGraw-Hill, 1992.

An excellent reference work for the layperson. Contains more than 7,000 signed articles. If this title is not available try the **Concise Encyclopedia of Science and Technology** by the same publisher.

123. McNeil, Ian, ed. **An Encyclopaedia of the History of Technology**. London: Routledge, 1990. 1,062 pp.

An essential reference.

124. Schapsmeier, Edward L., and Frederick H. Schapsmeier. **Encyclopedia of American Agricultural History.** Westport, CT: Greenwood Press, 1975. 467 pp.

Provides brief entries on terms, legislation, and people. May have limited use for the historian of agricultural science and technology.

125. Smith, Maryanna S. **Chronological Landmarks in American Agriculture**. Agricultural Information Bulletin, No. 425. Washington, DC: United States Department of Agriculture, Economics, Statistics, and Cooperatives Service, 1979. 103 pp.

Designed primarily for the layperson rather than the scholar. This source provides a listing of important developments in American agriculture, such as technology, crops, livestock and institutions. The listings often include a bibliographical citation.

126. Williams, Martha E., and Carolyn G. Robins. **Agricultural Databases Directory.** Bibliographies and Literature of Agriculture Series, No. 42. Beltsville, MD: National Agricultural Library, 1985. 175 pp.

Useful for an overview of the agricultural databases available by the mid-1980's. However, researchers should use the most recent edition of the **Directory of Online Databases** to verify current agricultural databases.

CHAPTER IV

SURVEYS

127. Anderson, James Richard. **A Geography of Agriculture in the United States Southeast**. Budapest: Akademiai Kiado, 1973. 135 pp.

    Describes the structure and patterns of farming and the changes that took place during the last century. Some discussion of technology and science but this study stresses economics.

128. Anthony, Kenneth R.M., et al. **Agricultural Change in Tropical Africa**. Ithaca: Cornell University Press. 1979. 326 pp.

    Addresses matters related to sub-Saharan agricultural innovations, productivity and research. Some discussion of technology, but the emphasis is economic and social interaction and change.

129. Bray, Francesca. **Agriculture**. Vol. 6, Part 2 of **Science and Civilization in China**. Cambridge: Cambridge University Press, 1984. 760 pp.

Provides a technological history of agriculture in China from 1400-1900, including information on implements, techniques and crops. In addition, this study attempts to compare and contrast agrarian change in Europe and China. A definitive and detailed work on the history of Chinese agricultural science and technology. Includes Chinese and Japanese as well as English language bibliography.

130. Boyce, James K. **Agrarian Impasse in Bengal: Institutional Constraints to Technological Change**. New York: Oxford University Press, 1987. 308 pp.

Explores the reasons for hunger in Bangladesh and West Bengal. Although this area has great soil fertility, agricultural production is extremely low. This study is concerned with the period from 1949 to 1980 in relation to technological, demographic and institutional change.

131. Bul'maga, L.E. "Nauchno-Tekhnicheskii Progress v Sel'skom Khoziastve Moldavskoi SSR." **Voprosy Istorii** [Soviet Union] I (1975): 23-31.

A review of scientific and technological progress in Soviet agriculture from 1959 to 1970. Emphasizes the Moldavian Soviet Socialist Republic.

132. Chambers, J.D., and G.E. Mingay. **The Agricultural Revolution, 1750-1880**. New York: Schocken Books, 1966. 222 pp.

A useful summary of British agriculture during the industrial revolution. Contends that mechanization and science did not aid farm production until the nineteenth century when demand began to exceed supply.

133. Cochrane, Willard Wesley. **The Development of American Agriculture: A Historical Analysis**. Minneapolis: University of Minnesota Press, 1979. 464 pp.

Primarily economic history, but this study includes a good, brief introduction to the development of technological hardware in the United States.

134. Cone, L. Winston, and J.F. Lipscomb. **The History of Kenya Agriculture**. Nairobi, Kenya: University Press of Africa, 1972. 160 pp.

Essentially an oral history of Kenyan agriculture. Of limited value for studies about Kenyan agricultural science and technology, but useful for understanding culture and rural problems.

135. Craig, G.M., ed. **The Agriculture of the Sudan**. Oxford: Oxford University Press, 1991. 468 pp.

Includes essays by twenty-one scholars and supplements J.D. Tothill's **Agriculture in the Sudan** (1948). Primarily economic and social in focus. Designed for the introductory student of Sudanese and African agricultural history.

136. Dhungana, Bhavani, ed. **Research, Productivity and Mechanization in Nepalese Agriculture: Seminar Report**. Kathmandu: Centre for Economic Development and Administration, Tribhuvan University, 1976. 308 pp.

Discusses agricultural research in Nepal. Compares traditional and mechanized farms and discusses agricultural production patterns. Economic in emphasis.

137. Fenton, A. **Scottish Country Life**. Edinburgh: John Donald Publishers, 1976. 255 pp.

A well-illustrated account of Scottish farming practices and implements, especially for tillage and harvesting. Emphasizes the eighteenth and nineteenth centuries.

138. **Fundamentals of Agricultural Production Techniques**. Arlington, VA: U.S. Joint Publications Research Service, 1974. 270 pp.

Translations from the Chinese monograph **Fundamentals of Agricultural Production Techniques** published by the Ministry of Agriculture of the People's Republic of China in 1965. The selections consider post-1949 Chinese agriculture in relation to science and technology.

139. Fussell, George E. **Farms, Farmers, and Society: Systems of Food Production and Population Numbers**. Lawrence, KS: Coronado Press, 1976. 332 pp.

A good overview of European agricultural practices with much about science and technology. Fussell's work is always worth reading.

140. Gras, Norman Scott Brien. **A History of Agriculture in Europe and America**. 2nd ed. New York: F.S. Crofts, 1946. Reprint. New York: Johnson Reprint Corp., 1968. 496 pp.

A description of some of the most important developments in the history of rural life in Europe and the United States. Topics include Roman agricultural history, the Medieval manor and peasant revolts as well as English and French farming practices and American land holding, animal husbandry and rural life until 1940.

141. Grigg, David B. **The Dynamics of Agricultural Change: The Historical Experience**. London: Hutchinson, 1982. 260 pp.

Some discussion of technological innovations and the role of transportation, but this study primarily concerns economic, social and demographic change.

142. ____. **English Agriculture: An Historical Perspective**. New York: Basil Blackwell, 1989. 256 pp.

Discusses the influence of the chemical revolution and improved technology on British agriculture.

143. Gupta, S.P. **Modern India and Progress in Science and Technology**. New Delhi: Vikas Publishing House, 1979. 164 pp.

A short but useful overview of the historical developments in Indian agricultural education and research during the twentieth century. Includes a list of specialized research centers and their date of establishment.

144. Heiser, Charles Bixler. **Seed to Civilization: The Story of Food.** Cambridge: Harvard University Press, 1990. 228 pp.

An ethnobiological history that discusses the origin of agriculture and the reasons people domesticated plants and animals. Emphasizes the contributions of plants to the food chain. Includes a discussion of breeding and ethical questions concerning present-day starvation and the depletion of the ecosystem.

145. Hill, R.D. **Agriculture in the Malaysian Region**. Budapest: Akademiai Kiado, 1982. 233 pp.

A survey of the economic aspects of Malaysian agriculture. Some discussion of crop distribution and rice and livestock production.

146. Hocking, Anthony. **South African Farming**. Cape Town: MacDonald South Africa, 1975. 64 pp.

A brief overview designed for the layman rather than the scholar.

147. Jabati, S.A. **Agriculture in Sierra Leone**. New York: Vantage Press, 1978. 349 pp.

An introduction to the agricultural activities of Sierra Leone. Not designed for the scholar, but given the lack of historical agricultural literature on this country, it merits consideration.

148. Kampp, Aage Hjalmar Hanse. **An Agricultural Geography of Denmark**. Budapest: Akademiai Kiado, 1975. 88 pp.

A brief history that emphasizes the cooperative movement and modern agricultural production. Stresses economic development.

149. Kostrowicke, Jerzy, and Roman Szczesny. **Polish Agriculture: Characteristics, Types and Regions**. Budapest: Akademiai Kiado, 1972. 120 pp.

A geographical survey of contemporary Polish agriculture. Includes a discussion of land use, farming practices, crops and livestock production. Introductory in nature.

150. Lombard, C. Stephen, and Alex H.C. Tweedie. **Agriculture in Zambia Since Independence**. Lusaka: Published by Neczam for the Institute of African Studies, University of Zambia, 1972. 105 pp.

An introduction to Zambian agriculture for the layman. Topics of discussion include crop and livestock production and marketing. Economic rather than technological or scientific in focus.

151. Masefield, Geoffrey Bussell. **A History of the Colonial Agricultural Service**. Oxford: Clarendon Press, 1972. 184 pp.

A history of the service from the eighteenth century through the 1950s. Includes a discussion of agricultural policy, research, extension and the contributions of individuals.

152. Medvedev, Zhores. **Soviet Agriculture**. New York: W.W. Norton, 1987. 464 pp.

Traces the transformation of the Soviet Union from an exporter to an importer of agricultural commodities. Covers the period from 1900 to 1986. Primarily economic and political in focus, but an essential study for anyone investigating Soviet agricultural history.

153. Minchinton, Walter E. **Essays in Agrarian History: Reprints, Edited for the British History Society**. 2 Vols. New York: A.M. Kelley, 1968.

Selected previously published essays covering English agriculture since 1750. Economic and social in emphasis, but some information about technology can be gleaned from this collection.

154. O'Reilly, Francis Dominic. **Thailand's Agriculture**. Budapest: Akademiai Kiado, 1983. 97 pp.

An overview of farming in Thailand. Discusses physical constraints, population, land use, crop and livestock production and agricultural regions. More economic than scientific or technological.

155. Paoletti, M.G., and G.G. Lorenzoni. "Agroecology Patterns in Northeastern Italy." In **Agriculture, Ecosystems and Environment**, 1-4, 139-54. Proceedings of an International Symposium on Agricultural Ecology and Environment, Padova, Italy, 5-7 April 1988.

An ecological history of the interrelationship between crops, weeds and pests. Some discussion of minimum and no tillage agricultural practices.

156. Rahman, Mushtaqur. **Agriculture in Pakistan**. Budapest: Akademiai Kiado, 1988. 150 pp.

Surveys agricultural conditions within a geographical framework. Discusses regions, weather, climate, land forms, technology, fertilizer and cropping patterns. A good introduction.

157. Randhawa, Mohindar Singh. **A History of Agriculture in India**. 4 Vols. New Delhi: Indian Council of Agricultural Research, 1980-1986.

A detailed narrative. Volume 1 covers the period to the twelfth century, while volume 2 discusses agriculture from the eighth to the eighteenth centuries. Volume 3 emphasizes the period from 1757 to 1947, and volume 4 discusses Indian agriculture to 1981.

158. Richards, Alan. **Egypt's Agricultural Development, 1800-1980**. Boulder: Westview Press, 1982. 292 pp.

A discussion of technological change, class structure and government policy. Describes the influences of Egypt's leaders, the large-scale land owners and international capitalism in the nation's agricultural development.

159. Schlebecker, John T. **Whereby We Thrive: A History of American Farming, 1607-1972**. Ames: Iowa State University Press, 1972. 342 pp.

A survey of American agriculture that emphasizes scientific and technological change. A good introduction.

160. Scott, Peter. **Australian Agriculture: Resource Development and Spatial Organization**. Budapest: Akademiai Kiado, 1991. 136 pp.

A contemporary survey by a geographer, who discusses resources, systems and potential. Useful for anyone coming to Australian agricultural history for the first time.

161. Shand, Richard Tregurtha, ed. **Technical Change in Asian Agriculture**. Canberra: Australian National University Press. 1973. 319 pp.

An analysis of technological change in grain production in India, Indonesia, the Philippines, Thailand and Myanmar (Burma) including the adoption of hybrid seeds and new hardware.

162. Shotskii, Vladimir Porfir'evich. **Agro-industrial Complexes and Types of Agriculture in Eastern Siberia**. Budapest: Akademiai Kiado, 1979. 130 pp.

A geographical introduction to farming methods and types. Contemporary rather than historical.

163. Skerman, P.J. "Agriculture in the Queensland Economy During the past 50 Years." **Journal of the Australian Institute of Agricultural Science** 53 no. 3 (1987): 185-91.

A brief survey that includes a description of soil and water conservation activities.

164. Smith, R.E.F. **Peasant Farming in Muscovy**. New York: Cambridge University Press, 1977. 289 pp.

This study examines tillage implements and livestock in addition to peasant farming in the Moscow, Toropets and Kazan regions.

165. Stratton, John M., and Jack Houghton Brown. **Agricultural Records in Britain, A.D. 220-1977**. 2nd ed. Hamden, CT: Archon Books, 1979. 259 pp.

Builds on the work of Thomas H. Baker entitled **Records of the Seasons, Prices of Agricultural Produce, and Phenomena Observed in the British Isles** (1883). An excellent source for common events and agricultural facts, such as weather and harvesting, that every historian of

English agriculture will find enlightening and useful.

166. Stross, Randall E. **The Stubborn Earth: American Agriculturalists on Chinese Soil, 1898-1937**. Berkeley: University of California Press, 1986. 272 pp.

Case histories of a dozen attempts to bring agricultural change to China. The subjects include plant collection, pest control, the improvement of cotton and wheat production, agricultural education and water conservation. Contends the American experts favored hardware solutions and met failure because they did not consider technological transfer in relation to Chinese culture.

167. Swaminathan, Monkombu Sambasivan. **Science and the Conquest of Hunger**. New Delhi: Concept Pub. Co., 1983. 508 pp.

Discusses the importance of science and technology in solving the problems of hunger and malnutrition, particularly regarding the conservation of plant and animal germplasm, plant breeding and crop research. See for agricultural trends in India and the tropics.

168. Troughton, Michael J. **Canadian Agriculture**. Budapest: Akademiai Kiado, 1982. 355 pp.

A geographical survey that emphasizes economic development, but a useful introduction.

169. Varjo, Uuno. **Finnish Farming: Typology and Economics**. Budapest: Akademiai Kiado, 1977. 145 pp.

Examines Finnish agriculture from the standpoint of the farm unit. Emphasizes farm population and income. Economic and statistical in focus rather than science and technology.

170. Volin, Lazar. **A Century of Russian Agriculture: From Alexander II to Krushchev**. Cambridge: Harvard University Press, 1970. 644 pp.

Divided into three parts: part 1 discusses peasants and landlords; part 2 analyzes the rise of communism; and part 3 describes collective agriculture and the influence of Krushchev. Parts 1 and 2 cover the period from 1850 to 1950.

171. Xu, Guohua, and Lynnette Jean Peel, eds. **The Agriculture of China**. Oxford: Oxford University Press, 1991. 300 pp.

An introduction to Chinese agriculture designed for business people and travellers. Not a study for scholars, although they may find it useful, if they are coming to the subject of Chinese agriculture for the first time.

CHAPTER V

AGRICULTURAL SCIENCE

172. Berthold, Rudolf. "Thunen im Lichte der DDR Forschung." **Jahrbuch fur Wirtschaftsgeschichte** [German Democratic Republic] 2 (1984): 81-91.

Critically reviews the work of Johann Heinrich von Thunen (1783-1850), who has been called the father of agricultural science in Germany.

173. Conner, E. H. Physicians and the Development of Scientific Agriculture, Empiricism to Science, 1731-1863. **Transactions & Studies of the College of Physicians of Philadelphia** 45 (October 1978): 316-35.

Contends that medicine was an empirical biological science until the nineteenth century, and that physicians became interested in agriculture as a manifestation of their scientific work. About 1830, however, science education began to shift from the medical schools because of the increasing complexity of each discipline. By 1863, agriculture had achieved the status of a separate science.

174. Conway, G.R. "The Properties of Agroecosystems." **Agricultural Systems** 24 no. 2 (1987): 95-117.

A general discussion of the ecology of agriculture.

175. Eyles, A.G., and D.G. Cameron. "The Contributions of Science to Australian Tropical Agriculture: Tropical Pasture Research." **Journal of the Australian Institute of Agricultural Science** 51 no. 1 (1985): 17-28.

Discusses the research on native and improved pastures as well as plant introductions and animal and plant nutrition.

176. Freeman, Orville L. "Scientific Agriculture: Keystone of Abundance." In **Science for Better Living: Yearbook of Agriculture 1968**, xvi-xlvi. Washington, D.C.: United States Department of Agriculture, 1968.

A sympathetic study that stresses the achievements of American agricultural science. See this yearbook for a general, popular introduction to the success stories in American agricultural science regarding a host of subjects, such as genetics, plant and livestock breeding and economic entomology.

177. Fried, M.I. "Historical Introduction to the Use of Nuclear Techniques for Food and Agriculture." **Bulletin, International Atomic Agency** 18 (Supplement 1976): 4-6.

An introductory note to a special issue dealing with the use of nuclear energy for food preservation and the improvement of livestock health. Emphasizes world-wide research.

178. Fussell, G.E. "Agricultural Science and Experiment in the Eighteenth Century: An Attempt at a Definition." **Agricultural History Review** 24, pt. 1 (1976): 44-47.

Survey of land holdings and grasslands agriculture in Europe. Fussell, however, does not develop the relationship of history and agricultural science as much as the title suggests, but he contends that demonstration plots are not scientific experiments. Only pure research can provide new agricultural knowledge.

179. _____. "History and Agricultural Science." **Agricultural History** 19 (April 1945): 126-127.

Contends that science could not be applied to British and European agriculture until the enclosure movement during the eighteenth century. Large-scale land owners broke from the land tenure systems of the past and, as John Humphries advised, began to use artificial manure and improved tillage practices to increase production and profits.

180. Fusonie, Alan, and Donna Jean Fusonie. **Twentieth Century Agricultural Science: Discovery, Use, Preservation**. Beltsville, MD: Associates of the National Agricultural Library, 1983. 252 pp.

     A collection of twenty-two symposium papers on various disciplines relating to the science of agriculture. Discusses accomplishments, trends and research.

181. Galli, R. "New Crops for Semi-Arid Regions of Mediterranean European Countries." **Fast Occasional Paper, No. 88, Forecasting and Assessment in Science and Technology**. Brussels: Directorate General for Science, Research and Development, Commission of the European Communities, 1986. 214 pp.

     Discusses the history of guayule and jojoba for the extraction of rubber. Concludes that it can be raised in most European Mediterranean countries but that it grows best in North Africa and the Middle East.

182. Gayko, D. "Geschichte der Agrarwissenschaften der DDR." **Jahrbuch fur Wirtschaftsgeschichte** no. 4 (1987): 229-32.

     Outlines the history of agricultural sciences in the German Democratic Republic to celebrate the 35th anniversary of the founding of the GDR Academy of Agricultural Science. Notes the beginnings of technical and scientific research during the 1950s.

183. Hall, Alfred Rupert. **The Scientific Revolution, 1500-1800**. London: Longman Group U.K., 1983. 373 pp.

Originally published as **The Scientific Revolution** in 1954, this study provides a thorough discussion of the main scientific developments and problems during the sixteenth and seventeenth centuries. Although agriculture is not discussed extensively, this book will provide the necessary intellectual foundation for work in this area and period.

184. Hartog, C. den. "Artsen, Landbouw, Voedselvoorziening en Voeding in de Periode van 1850-1950." **Voeding** 45 no. 4 (1984): 134-39.

Discusses the role of medical doctors in agricultural research to solve problems of health, nutrition and animal disease.

185. "Geschichte der Agrarwissenschaften der GDR." **Tagungbericht der Akademie der Landwirtschaftswissenschaften der Deutschen Democratischen Republik No. 254**, 1987. 192 pp.

Twenty-two symposium papers on the history of agricultural science in the German Democratic Republic. Discusses specific research activities, such as crop and livestock husbandry and agricultural engineering as well as activities at various universities. Includes a discussion of the transfer of research discoveries to the farm.

186. Jain, H.K. "Role of Research in Transforming Traditional Agriculture: An Emerging Perspective." **ISNAR Reprint Series, No. 4**, 1988. 15 pp.

The world population explosion during the 1960s transformed agriculture as nations attempted to use science and technology to improve food supplies. Genetic research substantially increased food supplies in Asia and Latin America. African agricultural improvement, however, depends on soil and water management as well as plant breeding.

187. Johnston, A.N. "Impact of Agricultural Science on Agriculture in New South Wales, 1935-1985--An Overview." **Journal of the Australian Institute of Agricultural Science** 53 no. 4 (1987): 247-53.

A brief, but useful introductory outline of agricultural science and development in New South Wales.

188. Kohlmeyer, Fred L.W., and Floyd L. Herum. "Science and Engineering in Agriculture: A Historical Perspective." **Technology and Culture** 2 (Fall 1961): 368-380.

Discusses the application of scientific knowledge to the technical problems in agriculture as well as the problems caused by insufficient capital. The authors also trace the development of agricultural engineering as a profession.

189. Kulikov, V.I. "Sovetskie Uchenye v Bor'be za Osvoenie Tselinnykh Zemel' I Povyshenie Effektivnosti Ikh Ispol'zonvaniia." **Istoriia SSR** [Soviet Union] no. 1 (1981): 20-36.

    Describes the efforts of Soviet scientists to increase grain production on virgin lands in Siberia and Central Asia from 1953 to 1978. During this period, production increased by 100 percent with the use of improved seeds and technology.

190. Landreth, Burnet. "The New Agriculture." **Proceedings of the American Philosophical Society** 45 (May-September 1906): 166-178.

    A potpourri discussion of the experiments that led to the development of steam traction engines and insecticides to eradicate the San Jose scale and boll weevil as well as to the application of electricity for agricultural purposes. These are incomplete generalizations that nevertheless indicate that American agriculture had taken a new scientific and technological direction.

191. Lechavlier, Hurbert A., and Morris Solotorovsky. **Three Centuries of Microbiology**. New York: McGraw Hill, 1965. 536 pp.

    A history of the origin and development of microbiology. Although this study primarily relates to humans see the section on soil microbiology for agricultural implications.

192. Mollah, W.S., and J.J. Daly. "Agriculture and Agricultural Science in the North-west, 1935-1985." **Journal of the Australian Institute of Agricultural Science** 53 no. 2 (1987): 68-77.

     Discusses the developments in climatology and crop and livestock production over a fifty-year period.

193. Rasmussen, Wayne D. "Scientific Agriculture." In **Technology in Western Civilization**, edited by Melvin Kranzberg and Carroll W. Pursell Jr., Vol. 2, 337-53. New York: Oxford University Press, 1967.

     Reviews the major scientific changes during the twentieth century in American agriculture. Brief sections on genetics, hybridization of corn and wheat and the development of commercial fertilizers.

194. Russell, Edward J. **A History of Agricultural Science in Great Britain, 1620-1954**. London: Allen and Unwin, 1966. 493 pp.

     A general discussion of the major scientists and their discoveries during three centuries of British agriculture. Summarizes the history of agricultural research institutions, including the Rothansted Experiment Station.

195. Saltini, Antonio. **Storia delle Scienze Agrarie**. 4 Vols. 2nd ed. Bologna: Edagricole, 1984-1989.

The history of agricultural science in Europe from the Renaissance to the twentieth century. Emphasizes Italy.

196. Sims, H.J. "Agriculture and Agricultural Science in the Past Fifty Years: Victoria." **Journal of the Australian Institute of Agricultural Science** 53 no. 4 (1987): 238-46.

A brief review of agricultural science, production and economics in the state of Victoria.

197. Slocum, Walter L. "Careers of Ph.D.'s in the Agricultural Sciences." **Agricultural Science Review** 6 no. 1 (1968): 18-23.

Surveys employment possibilities in agricultural science. Good for historical comparisons about job mobility, earning potential and changes of field.

198. Spurling, M.B. "Agricultural Achievements in South Australia." **Journal of the Australian Institute of Agricultural Science** 53 no. 2 (1987): 61-67.

Briefly describes the developments in crop production, irrigation, livestock raising and viticulture.

199. Stoyanov, K.H. "Developments and Achievements of Agricultural Science in the Field of Mechanization of Agriculture in the last 25 Years." **Selskostopanska Tekhnika** [Bulgaria] 26 no. 1 (1989): 5-27.

Discusses the developments in mechanization for sowing, fertilizing and cultivating crops. New inventions and the testing of machines are noted. Topics include tractors and forage, vegetable and fruit production.

200. Tiver, N.S. "Desert Conquest," **Agricultural Science** no. 1 (1988): 5, 12-17.

Surveys the transformation of the Ninety-Mile Desert in Australia to the farmlands known as the Coonalpyn Downs. During the 1930s, agricultural scientists began using lucerne to prevent erosion and improve soil fertility. See for the history of land development in this region.

201. Wells, George S. **Garden in the West: A Dramatic Account of Science in Agriculture**. New York: Dodd, Mead, 1969. 270 pp.

A personal reflection about the state of agricultural science, particularly in relation to pests. Suggests possibilities for future work.

202. Wik, Reynold Millard. "Henry Ford's Science and Technology for Rural America." **Technology and Culture** 3 (Summer 1962): 247-258.

A brief history of Henry Ford's interest in agriculture, particularly in relation to agricultural experiments on test farms and the application of chemistry to agriculture.

203. _____. "Science and American Agriculture." In **Science and Society in the United States**, edited by David D. Van Tassel and M.G. Hall, 81-106. Homewood, IL: Dorsey Press, 1966.

Discusses the progress of agricultural science since the colonial period. Wik also notes the early involvement of the federal government in agriculture and the scientific work of the United States Department of Agriculture.

204. Wilmot, Sarah. **The Business of Improvement: Agriculture and Scientific Culture in Britain, c.1700-c.1870**. Bristol, United Kingdom: University of Bristol, 1990. 130 pp.

Contends that agricultural societies conducted legitimate scientific experiments even though their influence was limited. The agricultural elite supported these societies as a means to reorganize their estates and gain recognition for supporting progressive change. These arguments are often asserted rather than proved, but this study is a good introduction to British agricultural science.

205. Wilson, M.L. "Survey of Scientific Agriculture." **Proceedings of the American Philosophical Society** 86 (1943): 52-62.

A brief overview of the agricultural concerns of the American Philosophical Society published in its **Proceedings** and **Transactions** from 1768-1809, including Jefferson's experiments with a new moldboard plow and early agricultural education. The discussion of the Society's activities is the most important aspect of this study.

206. Ziegler, G.M. "Agricultural Magic." **Scientific Monthly** 27 (July 1928): 69-76.

The Romans and early European farmers explained complicated procedures as the result of magic. Science would replace superstition only after centuries of scientific experimentation and education. Useful for background on ancient agricultural superstitions and the attempts to explain the unknown.

CHAPTER VI

WOMEN

207. Baser, Heather. **Technology, Women and Farming Systems**. Studies in Technology and Social Change Series, No. 5. Ames: Iowa State University, Technology and Social Change Program, 1988. 26 pp.

    In Africa, women primarily are responsible for farming. Extension organizations, however, tend to emphasize technological transfer for men and leave women outside the training and educational programs, particularly for the improvement of subsistence agriculture.

208. Calhoun, M.L., and K.A. Houpt. "Women in Veterinary Medicine." **Cornell Veterinarian** 66 (October 1976): 455-475.

    A history of women in veterinary medicine during the twentieth century. International focus. See for a listing of women in the profession.

209. Creevey, Lucy E., ed. **Women Farmers in Africa: Rural Development in Mali and the Sahel**. Syracuse: Syracuse University Press, 1986. 212 pp.

Discusses the roles of women farmers and provides case studies of specific projects and programs designed to improve agricultural development.

210. Faruqui, A.M., Mohamed Hassan, and Gabriella Sandri. **The Role of Women in the Development of Science and Technology in the Third World: Proceedings of the Conference Organized by the Canadian International Development Agency and the Third World Academy of Sciences, ICTP, Trieste, Italy, 3-7 October 1988**. Teaneck, NJ: World Scientific, 1991. 970 pp.

An excellent introduction. See the bibliographical references for additional citations.

211. Fera, Darla. **Women in American Agriculture, A Selected Bibliography**. National Agricultural Library List No. 103. Beltsville, MD: National Agricultural Library, 1977. 30 pp.

A bibliography with brief annotations to works available in the National Agricultural Library in Beltsville, Maryland and the Library of Congress. Although the primary focus is social and economic history, some citations deal with women in relation to agricultural science and technology.

212. Ferguson, Anne E., et al. **Resource Guide, Women in Agriculture**. East Lansing, MI: Bean/Cowpea Collaborative Research Program, 1984-.

A series of resource guides that have been prepared by the staff of the Bean/Cowpea Collaborative Research Support Program at Michigan State University. Guides have been compiled for Botswana, Cameroon and Guatemala. An extremely useful, annotated series for agricultural historians examining the role of women in agricultural technology.

213. Fernandez, Maria E. "Technological Domains of Women in Mixed Farming Systems of Andean Peasant Communities." In **Gender Issues in Farming Systems Research and Extension**, edited by Susan Poats, Marianne Schmink and Rita Spring, 213-21. Boulder: Westview Press, 1988.

Discusses the importance of gender in the mixed farming systems of the Andes. Women are the principal herders; therefore, extension specialists must locate those women when creating development programs that involve scientific and technological change.

214. Gates, Jane Potter. **Women in Agriculture, January 1979--July 1991.** Quick Bibliography Series QB91-150. Beltsville, MD: National Agricultural Library, 1991. 42 pp.

Excellent source for up-to-date bibliography on the general topic of women in agriculture. The citations are from the **AGRICOLA** database and therefore emphasize the United States. Alphabetical order by title, with author index. No annotations and limited to English publications. Use **AGRICOLA** to update.

215. Henderson, J.L., and B.E. Cooper. "The Representation of Women Scientists in Land-Grant Colleges of Agriculture." **NACTA Journal** 31 (June 1987): 14-17.

See for statistics concerning the number of women employed in the agricultural sciences, including academic disciplines, in the United States.

216. Jain, S.C., ed. **Women and Technology.** Jaipur, India: Rawat, 1985. 200 pp.

Includes sixteen chapters most of which emphasize women and agricultural technology in India.

217. Jensen, Joan M. **With These Hands: Women Working on the Land**. Old Westbury, NY: Feminist Press, 1981. 295 pp.

Wide-ranging reflections by American women about their agricultural work. Includes useful information concerning the use of technology, such as Pueblo grinding tools.

218. Kaur, Malkit. **Rural Women and Technological Advancement**. Delhi: Discovery Publishing House, 1988. 213 pp.

This study emphasizes technological adoption by women in India. Some social analysis. Good bibliography for additional research.

219. Kessler, Shelly. **Third World Women in Agriculture: An Annotated Bibliography**. New York: National Council for Research on Women, 1985. 71 pp.

An important starting point for the history of women in agriculture. This well-annotated bibliography includes citations to "fugitive literature," such as unpublished papers from conferences and in-house publications. Collections in the New York City and Washington D.C. areas were utilized. Topical arrangement with a section on technology.

220. Kumar, Shubh K. "Women's Role and Agricultural Technology." In **Accelerating Food Production in Sub-Saharan Africa**, edited by John W. Mellor, Christopher L. Delagado and Malcolm J. Blackie, 135-47. Baltimore: Johns Hopkins University Press, 1987.

Ensuring women farmers access to new information and technologies is a special challenge. Discusses the need for development of technological "packages" that address constraints on women's labor because women farmers are often trying to deal with household as well as farm responsibilities.

221. Ogilvie, Marilyn Bailey. **Women in Science: Antiquity Through the Nineteenth Century: A Biographical Dictionary with Annotated Bibliography**. Amherst, MA: MIT Press, 1986. 254 pp.

Biographical accounts of women in science from antiquity to the early twentieth century, some of whom can be loosely associated with agriculture, e.g. botany, chemistry and horticulture.

222. Rossiter, Margaret W. "Women and the History of Scientific Communication." **Journal of Library History** 21 (1986): 39-59.

Women have been involved in the sciences much longer than many historians realize, because they primarily have been associated with matters of communication rather than creation. Contends that as communication becomes more important, the prestige of women in the sciences will increase.

223. _____. "'Women's Work' in Science, 1880-1910." **Isis** 71 (1980): 381-398.

During the late nineteenth century, women who completed work in higher education were allowed the opportunity to use their special training in scientific fields. Most women, however, served as assistants or in the field of home economics.

224. _____, Barbara J. Harris, and Suzanne Hildenbrand. **Women Scientists in America: Struggles and Strategies to 1940**. Baltimore: Johns Hopkins University Press, 1982. 439 pp.

A comprehensive study of the history of American women scientists. The course of women breaking into the male-dominated ranks of science is examined in detail.

225. Sachs, Carolyn, E. **The Invisible Farmers: Women in Agricultural Production**. Totowa, NJ: Rowman & Allanheld, 1983. 153 pp.

Primarily a social and economic analysis of women in American agriculture, but it includes some discussion of technology. A good introduction if used with care, because the author over emphasizes the oppression of women farmers by their husbands.

226. Sutherland, Alistair J. "The Gender Factor and Technology Options for Zambia's Subsistence Farming Systems." In **Gender Issues in Farming Systems Research and Extension**, edited by Susan Poats, Marianna Schmink and Anita Spring, 389-406. Boulder: Westview Press, 1988.

Discusses the importance of gender in farm systems research. Researchers often underestimate the proportion of female-heads of households and do not "seek" the opinion of women farmers when designing rural development programs.

227. Todhunter, Elizabeth Neige. "Women in Nutrition." **Professional Nutritionist** 9 (Fall 1977): 12-14.

A brief discussion of the professional problems that women encountered in the field of nutrition science and the ways in which they began to solve those difficulties. Includes brief profiles of women leaders in this field.

228. Whitehead, Vivian B. **Women in American Farming: A List of References**. Davis, CA: Agricultural History Center, University of California, Davis, 1987. 104 pp.

An excellent bibliography about women in agriculture in the United States. Some of the citations include scientific and technological subjects. A good place to start.

229. **Women and Agricultural Technology: Relevance for Research. Report From CGIAR Inter-Center Seminar on Women and Agricultural Technology, Bellagio, Italy, 25-29 March 1985**. 2 Vols. New York: Rockefeller Foundation and the International Service for National Agricultural Research, The Hague, Netherlands, 1985.

A report from a conference held to review the special needs of women farmers, especially in Africa, Asia and Latin America. In areas of technological transfer, gender must be considered to ensure that women farmers benefit from it.

230. **Women in Rice Farming: Proceedings of a Conference on Women in Rice Farming Systems, Manila, Philippines, 26-30 September 1983**. Hants, England: Gower, 1985. 531 pp.

A collection of papers discussing the role of Asian and African women in rice cultivation and post-harvest processing. Also provides information on the effect of gender on technology transfer. A number of the papers are country specific, such as Bangladesh, India, Japan, Indonesia and Nepal.

231. Zarkovic, Milica. "Female Labor in the High Growth Agricultural Regions of India." **Peasant Studies** 14 no. 2 (1987): 81-103.

This study traces the influence of women on the Green Revolution in the northern Indian states of Punjab and Haryana between 1961 and 1981.

CHAPTER VII

AGRONOMY

232. Aldrich, Daniel G. Jr. "Some Personal Reflections on Soil Science and Agriculture, 1936 to 1986." **Soil Science Society of America Journal** 51 (November/December 1987): 1401-1405.

     Discusses the changes in soil science over a fifty year period. Argues that as American agriculture moves from domestic commercial emphasis to an internationally interdependent system for the distribution of food and fiber, scientific research must address societal concerns, such as the environment, health and nutrition.

233. Boulaine, J. **Histoire de Pedologues et de la Science des Sols**. Paris: INRA, 1989. 298 p.

     A history of soil science with emphasis on agricultural chemistry during the eighteenth and nineteenth centuries. Includes developments since 1950 and outlines problems to be solved.

234. Brown, P.E. "The Beginnings and Development of Soil Microbiology in the United States." **Soil Science** 40 (July-December 1935): 49-58.

Traces the beginning of soil science as a discipline to 1900. Evaluates the work of George T. Moore, Jacob G. Lipman, Frederick D. Chester, F.L. Stevens, W.A. Withers, and Edward B. Voorhees.

235. Carter, Vernon Gill, and Tom Dale. **Topsoil and Civilization**. Norman: University of Oklahoma Press, 1974. 292 pp.

A synthesis of the relationship between soil and civilization. Contends that civilizations have risen and fallen in direct proportion to their treatment of the soil.

236. Cline, M.G. "Historical Highlights in Soil Genesis, Morphology, and Classification." **Soil Science Society of America Journal** 41 (March-April 1977): 250-254.

A brief look at various pioneers in soil science and the development of the Soil Survey and Soil Science Society of America. Includes profiles of Eugene Walderman Hilgard, Milton Whitney, G.N. Coffey and C.F. Marbut.

237. De 'Sigmond, Alexius A.J. "Development of Soil Science." **Soil Science** 40 (1935): 77-86.

Discusses the development of soil science from the eighteenth century, when it was based on appearance, to the early twentieth century, when it became a separate scientific discipline.

238. Fisher, Ora S. **Ten Years of Agronomy Extension, 1915-1924**. United States Department of Agriculture, Circular No. 22, 1928. 24 pp.

Discusses the work of the agronomy specialist regarding new demonstration techniques, smut control, seed and soil improvement and the use of fertilizers.

239. Gardner, W.H. "Historical Highlights in American Soil Physics, 1776-1976." **Soil Science Society of America Journal** 41 (March-April 1977): 221-229.

Traces the beginning of American soil science to Thomas Jefferson. Contends that he was the first practicing soil physicist. Notes the application of chemistry and physics to the study of soils during the eighteenth and nineteenth centuries as well as the beginnings of professional journals and organizations. Brief and impressionistic.

240. Gelburd, Diane E. "Managing Salinity: Lessons From the Past." **Journal of Soil and Water Conservation** 40 (July/August 1985): 329-331.

A discussion of land salinity from irrigation as a cause of the decline of Mesopotamian civilization. Gelburd also contends that California faces increasing salinization of its agricultural land at a cost of $80 million in lost productivity annually.

241. **Glossary of Soil Science Terms**. Ottawa: Research Branch, Canada Department of Agriculture, 1976. 44 pp.

A brief dictionary.

242. Greenwood, L. Larry. **KWIC Index to the Commonwealth Bureau of Soils Annotated Bibliographies of Soils and Fertilizers, 1956-1972**. Manhattan, KS: Bibliography Series, No. 13, Kansas State University, 1974. 198 pp.

A dated but useful index for the study of agronomy.

243. Harlan, Jack R. **Crops & Man**. Madison: American Society of Agronomy, Crop Science Society of America, 1992. 284 pp.

An introductory synthesis of the history of crops and agriculture.

244. Held, R. Burnell, and Marion Clawson. **Soil Conservation in Perspective**. Baltimore: Published for Resources for the Future by the Johns Hopkins Press, 1965. 344 pp.

Discusses the success and obstacles of the conservation movement from the 1930s to the 1960s. Emphasizes economics and social and political relationships rather than a historical account of the application of science and technology to soil conservation.

245. Hulbert, Archer B. **Soil: Its Influence on the History of the United States**. New Haven: Yale University Press, 1930. 227 pp.

A study of the influences of soil on American expansion from the colonial period to the settlement of the West Coast. Also discusses climatic, geological and hydrographic features.

246. Jacks, G.V., R. Tavernier, and D.H. Bolach. **Multilingual Vocabulary of Soil Science**. Rome: Land & Water Division, Food and Agriculture Organization of the United Nations, 1960. 28 pp.

A useful dictionary with text in English, French, Spanish, German, Portuguese, Italian, Dutch, Swedish and Russian.

247. Jenny, Hans. **E.W. Hilgard and the Birth of Modern Soil Science**. Pisa, Italy: Agrochimica, Instituto di Chimica Agraria dell'Universita Pisa 1961. 144 pp.

A biography of America's foremost soil scientist, whose work was both theoretical and applied. Hilgard published important studies on the soils in the cotton states and California. Contends that Hilgard discovered the significance of climate in the formation of the soil.

248. Jerzak, M. "Agronomia Spoleczna Jej Rola w Nauce i Praktyce Rolniczej." **Wies Wspolczesna** 27 no. 2 (1983): 42-47.

Brief history of agronomy in Poland. Notes farmers' organizations and increased professionalization of agricultural advisors. Urges free advice by agronomists to help farmers produce for a profit.

249. Karlen, D.L., et al. "Soil Tilth: A Review of Past Perceptions and Future Needs." **Soil Science Society of America Journal** 54 (January-February 1990): 153-61.

Reviews the history and literature of soil tilth. Argues the current definition of soil tilth is too restrictive and suggests that it be defined as a "tilth-forming process" to include soil and water management problems, erosion, productivity and long-term sustainability as factors in its formation.

250. Kellogg, Charles E. **Agricultural Development--Soil, Food, People, Work**. Madison, WI: Soil Science Society of America, 1975. 233 pp.

Traces, in part, the development of soil science in developing nations. Provides examples of practices in developed nations that can be used to improve agriculture in underdeveloped countries.

251. ____. "The Soil Survey." In **After a Hundred Years: Yearbook of Agriculture, 1962**, 187-201. Washington, D.C.: United States Department of Agriculture, 1962.

Discusses the study of soil problems since the 1890s, when Americans first became interested in soil science.

Dates the beginning of soil mapping to 1899 and analyzes soil testing as an aid to farmers. Briefly discusses the use of fertilizer and conservation methods.

252. _____ . "We Seek; We Learn." In **Soil: Yearbook of Agriculture, 1957**, 1-11. Washington, D.C.: United States Department of Agriculture, 1957.

A discussion of the major advances in soil science. Generalities rather than specific examples, but a place to start.

253. Komornicki, T. "Historia Polskiego Towarzystwa Gleboznawczego w Piecdziesieciolecie Powstania." **Roczniki Gleboznaucze** 29 no. 2 (1988): 5-42.

A history of the Polish Society of Soil Science with emphasis on achievements during its first fifty years.

254. Krupenikov, I.A. **Istoriya Pochvovedenuya ot Vremeni Yego Zarozhdeniya do Nashikh Dney**. Moscow: Nauka Press, 1981. 327 pp.

A general history of soil science from its beginning to the present.

255. **Life and Work of C.F. Marbut, Soil Scientist: A Memorial Volume**. Columbia, MO: Soil Science Society of America, 1942. 271 pp.

The historian will find the bibliography of Marbut's papers as well as the eleven complete papers and reviews of other publications to be the most useful parts of this remembrance by his friends.

256. Lipman, Jacob G. "A Quarter Century Progress in Soil Science." **Journal of the American Society of Agronomy** 25 (January 1933): 9-25.

A review of the major developments in soil science. The article is topically arranged to include matters relating to soil surveys in the United States and Europe, soil chemistry and microbiology, plant nutrition, soil erosion and field plots.

257. Lutz, J.F. "History of the Soil Science Society of America." **Soil Science Society of America Journal** 41 (January-February 1977): 152-173.

Discusses the organizational history of the association, including the groups involved, and its emergence as a separate professional, scientific body.

258. MacDonald, Angus H. **Early American Soil Conservationists**. United States Department of Agriculture, Miscellaneous Publication, No. 449, 1941. 63 pp.

Profiles eight early soil conservationists--Jared Eliot, Samuel Deane, Solomon Drown, John Taylor, John Lorain, Isaac Hill, Nicholas Sorsby and Edmund Ruffin--who were active during the eighteenth and nineteenth centuries. Emphasizes soil erosion and control.

259. Maclean, J.T. "Soil Science History, January 1979-February 1986." **Quick Bibliography Series of the National Agricultural Library**. Beltsville, MD: National Agricultural Library, 1986. 16 pp.

 A good introduction to the recent literature on the history of soil science and management.

260. McCall, A.G. "The Development of Soil Science." **Agricultural History** 5 (April 1931): 43-56.

 A brief, historical sketch of international soil science. Concentration on advances in the United States, Europe and the Soviet Union. Notes the work of van Helmont, Liebig, Gilbert, de Saussures, Sprengel, Thaer, Dokuchaiev and Whitney.

261. Olson, Lois. "Columella and the Beginning of Soil Science." **Agricultural History** 17 (April 1943): 65-72.

 Discusses the work of this Spanish-born, first-century Roman who developed new agricultural techniques, such as terracing, and who recommended planting crops that returned nutrients to the soil.

262. _____, and Helen L. Eddy. "Pietro de Crescenziz: The Founding of Modern Agronomy." **Agricultural History** 18 (January 1944): 35-40.

 This Sicilian judge (1233-1320) studied agricultural techniques and advocated crop rotation and soil conservation. He also supported the use of manure to improve soil fertility. He authored **Opus Ruralium Commodorum** in which he discussed all phases of farming.

263. Plaisance, Georges. **Dictionnaire des Sols**. New Delhi: Published for the United States Department of Agriculture and the National Science Foundation by Amerind Pub. Co., 1981. 1,109 pp.

A useful dictionary for terminology relating to soils.

264. Pondel, H. "Bibliografia Publikacji Polskich z Zakresu Gleboznawstwa, Chemii Rolnej, Uprawy Roli i Gospodarki Wodnej. Rok 1985." **Roczniki Gleboznaucze** 29 no. 1 (1988): 153-201.

A bibliography of Polish sources for the study of soil science, tillage, and agricultural chemistry and technology.

265. Sandor, J.A., and N.S. Eash. "Significance of Ancient Agricultural Soils for Long-Term Agronomic Studies and Sustainable Agriculture Research." **Agronomy Journal** 83 no. 1 (1991): 29-37.

Ancient agricultural sites enable the study of soils in relation to cultivation. These sites help agronomists understand soil tilth, fertility and conservation as well as sustainability over an extended period of time. Peru and Mexico are used a examples.

266. Shaw, Byron T., ed. **Soil Physical Conditions and Plant Growth**. New York: Academic Press, 1952. 491 pp.

A critical evaluation of the relationship between the physical conditions of the soil and plant growth.

267. Smith, G.R. "Pioneering on the Conservation Frontier." **Soil Conservation** 42 (December 1976): 12-13.

Discusses the technological methods used to prevent soil erosion, such as building terraces and ponds and reseeding eroded crop lands.

268. Strzemski, M. "Czterysta Piecdziesiat lat Gleboznawstwa Polskiego." **Roczniki-Gleboznawcze** 29 no. 2 (1988): 43-73.

An overview of four hundred and fifty years of soil science in Poland.

269. Swanson, Louis. "A Century of Periodic Research on Soil Conservation." In **The Agricultural Scientific Enterprise: A System in Transition**, edited by Lawrence Busch and William B. Lacy, 131-147. Boulder: Westview Press, 1986.

Contends that when farmers operate on the edge of profitability, they will invest needed capital for long term soil conservation measures.

270. Thomas, G.W. "Historical Developments in Soil Chemistry: Ion Exchange." **Soil Science Society of America Journal** 41 (March-April 1977): 230-238.

Discusses the early ideas about ion exchange during the last half of the nineteenth century. A detailed look at the most important contributions to the understanding of ion exchange in soils. Chronological in organization. Emphasizes American and European scientists.

271. Throckmorton, R.I. "History of the American Society of Agronomy." **Journal of the American Society of Agronomy** 33 (December 1941): 1135-40.

Lists past officers and editors as well as the fellows of the Society from 1908-1940.

272. Varallyay, G. "Soil Mapping in Hungary." **Agrokemia es Talajtan** 38 no. 3-4 (1989): 695-714.

A survey of soil mapping in Hungary from 1858 to the present. Discusses problems that have been encountered and the new approaches used for modern, large-scale soil mapping.

273. Villiers, J.M. de. "Highlights of Soil Chemistry Research in Southern Africa, 1953-1978." **Technical Communications, No. 165**, 71-82. Department of Agricultural Technical Services, Republic of South Africa, 1980.

Agricultural research has emphasized the importance of potassium while placing less stress on nitrogen. The chemistry of micronutrients, heavy metals and salt-affected soils also have been areas of study.

274. Waksman, Selman A. "The Background of Soil Science." **Soil Science** 101 (January 1966): 6-10.

A brief discussion of the contributions of George H. Cook, Jacob G. Lipman, Edward B. Voorhees and William H. Martin to soil science.

275. Warkentin, B.P. "Soil Science for Environmental Quality--How Do We Know What We Know?" **Journal of Environmental Quality** 21 (April/June 1992): 163-66.

Discusses the last 150 years by dividing that period into four sections to analyze the questions that soil scientists asked. A good introduction to the ideas that have shaped agronomy.

276. Wells, C.B., and J.A. Prescott. "Origins and Early Development of Soil Science in Australia." In **Soils: An Australian Viewpoint**, 3-12. Melbourne: Council for Scientific and Industrial Research, 1983.

Traces the development of agronomy in Australia to the depletion of the soil during the late eighteenth and early nineteenth centuries. Soil testing at the turn of the twentieth century led to an increase in the use of fertilizers, and soil science education began at the universities.

277. Whitney, Milton. **Soil and Civilization: A Modern Concept of the Soil and the Historical Development of Agriculture**. New York: D. Van Nostrand Co., 1925. 278 pp.

A non-technical discussion of soil science for the educated reader. Includes a description of the important soils in the United Sates as well as methods of soil control. Dated but useful.

278. _____ . "Soil Investigations in the United States." In **Yearbook of Agriculture, 1899**, 335-46. Washington, D.C.: United States Department of Agriculture, 1899.

An overview of the importance of chemical analysis and research to solving soil problems. Notes the significance of bacteriological research and discusses the effects of irrigation. Stresses the importance of soil investigations and the utility of such studies.

279. Wild, Alan. **Russell's Soil Conditions and Plant Growth**. 11th ed. New York: John Wiley & Sons, 1988. 991 pp.

Discusses Sir Edward J. Russell's research during the late nineteenth and early twentieth centuries which led to the discovery of the "principle" of plant nutrition. Traces the foundations of plant physiology and science and the development of soil bacteriology from 1800 to 1860.

280. Yaalon, D.H. "Forerunners and Founders of Pedology as a Science." **Soil Science** 147 no. 3 (1989): 225-226.

     The publications on soils by W. Cobbett (1830), J. Morton (1838), and C.S. Springel (1837) provide the foundation for pedology or soil science.

281. Zonn, S.V. "The Present Day State and Prospects of Soil Science Development." **Nauchnye Doklady Vysshei Shkoly BiologicheskieNauki Moscou** 1 (1989): 17-27.

     A brief history of soil science in the Soviet Union. Proposes the organization of an international body to coordinate research on the control and protection of the soil.

CHAPTER VIII

METEOROLOGY

282. Dunlap, Thomas R. "Agricultural Research and the Concept of Climate." **Agricultural History** 63 (Spring 1989): 152-161.

    During the late nineteenth century, research on climate became the responsibility of the United States Department of Agriculture, land-grant colleges and the agricultural experiment stations. Agricultural scientists now began to stress meteorological research for the basis of their recommendations regarding the use of climate to increase crop and livestock production.

283. Feldman, Theodore Sherman. "The History of Meteorology, 1750-1800: A Study in the Quantification of Experimental Physics." Ph. D. diss., University of California, Berkeley, 1983. 297 pp.

    At the end of the eighteenth century, physics changed from a qualitative pursuit to a quantitative science based on mathematical theory and experimentation. After 1770 the study of weather changed from subjective experience by the individual to objective, scientific analysis by meteorological societies that collected data for rigorous mathematical analysis.

284. Fleming, James Rodger. **Meteorology in America, 1800-1870**. Baltimore: Johns Hopkins University Press, 1990. 264 pp.

A standard history. This is the place to begin research on meteorology in the United States.

285. Frisinger, H. Howard. **The History of Meteorology to 1800**. New York: Science History, 1977. 148 pp.

This history begins at 600 B.C., when meteorological authority was based on Aristotle's **Meteorologica**. Argues that scientific meteorology became established during the seventeenth and eighteenth centuries with the development of special instruments, such as the thermometer, barometer and hygrometer and systematic meteorological observation.

286. Fusonie, Alan, and William Hauser. "Climate History at the National Agricultural Library." **Agricultural History** 63 (Spring 1989): 36-50.

The National Agricultural Library in Beltsville, Maryland, holds materials relating to weather that date from the Roman period to the twentieth century.

This article discusses those print and non-print materials.

287. Halacy, Daniel S. **The Weather Changers**. New York: Harper & Row Publishers, 1968. 246 pp.

Discusses the methods used for the artificial production of rain, the suppression of fog and the prediction of weather.

288. **History of Climatological Publications: A History of Publications Containing Climatological Data**. Washington, D.C.: U.S. Weather Bureau, 1959. 34 pp.

A dated but useful introductory source.

289. Hughes, Patrick. **A Century of Weather Service: A History of the Birth and Growth of the National Weather Service, 1870-1970**. New York: Gordon and Breach, 1970. 212 pp.

Discusses the establishment and early development of the Weather Bureau, including the adoption of balloons, airplanes and satellites.

290. Jing, K. "A Study on the Relationship Between Soil Erosion and the Geographical Environment in the Middle Yellow River Basin." **Chinese Journal of Arid Land Research** 1 no. 4 (1988): 289-99.

Describes the history of climate and soil erosion in this area of China. Emphasis placed on soil structure and vegetative cover.

291. Jones, E. L. **Seasons and Prices: The Role of Weather in English Agricultural History**. London: George Allen & Unwin, Ltd., 1964. 193 pp.

A general study of the effects of the weather and seasons on English agricultural history. Emphasizes the influence of the weather on farm income and production rather than the effects of science and technology. Covers the period 1728-1911.

292. Koelsch, William A. "Ben Franklin's Heir: Alexander McAdie and the Experimental Analysis and Forecasting of New England Storms, 1884-1892." **New England Quarterly** 59 no. 4 (1986): 523-543.

Alexander George McAdie (1861-1943), a British immigrant and Harvard-trained scientist, pursued meteorological research in the U.S. Signal Corps, particularly regarding atmospheric electricity. He continued his research upon leaving the Army and popularized his work by writing for newspapers and magazines.

293. Mani, A. "History of Rainfall Measurement." In **Water for Human Needs: Proceedings of the 2nd World Congress on Water Resources**, Vol. 3, **Development & Meteorology**, 393-402, December, 1975.

Discusses the development of precipitation gauges during the past two thousand years as well as the efforts at standardization during the last century to reduce errors in measurement. Argues that meteorological radar can eliminate errors in gauging but such a radar and computer systems that can function independently from gauges have not yet been developed.

294. Miller, Eric R. "American Pioneers in Meteorology." **Monthly Weather Review** 61 (July 1933): 189-193.

A brief discussion of the development of the U.S. Weather Service from its origin in the Medical Department of the Army in 1814 through the 1840s. Emphasis given to the work of Simeon DeWitt, Maxine Bocher, George Hadley, William Redfield, James Espy, Joseph Henry, Albert Myer and William Ferrel. Includes background information on the development of weather mapping and the founding of the national weather forecasting service.

295. Oliver, J. "Problem of Agri-Climatic Relationships in Wales in the Eighteenth Century." In **Weather and Agriculture**, edited by James A. Taylor, 187-200. New York: Oxford University Press, 1967.

Argues that any assessment of the effects of weather and climate on the agriculture of eighteenth century Wales is difficult because instrumental or descriptive data is lacking. Only two instrumental records exist. Those records, kept by David Pennant and Dr. Perrott Williams, indicate temperature, wind direction and general weather patterns.

296. Roe, Frank G. "The Alberta Wet Cycle of 1899-1903: A Climatic Interlude." **Agricultural History** 28 (July 1954): 112-120.

A study of the Lacombe District during the most pronounced wet cycle in Alberta's history.

297. Schwartz, Leonard E. "Artificial Weather Modification: A Case Study in Science Policy." **Technology and Society** 5 (July 1969): 44-48.

Discusses the cloud seeding activities of Vincent Schaefer with dry ice in December 1946, which produced a hard snow storm. Surveys the activities of corporate enterprise and the federal government in rain-making ventures.

298. Shaw, Lawrence H., and Donald D. Dorset. "The Effects of Weather and Technology on the Corn Yields in the Corn Belt, 1929-1962." United States Department of Agriculture, **Agricultural Economic Report, No. 80**, 1965. 39 pp.

Contends that the effect of the weather on corn yields has been insignificant. Hybrid seeds, nitrogen fertilizer, pesticides and increased planting rates have improved and stabilized yields in both good and bad weather. Argues that technology rather than weather is the most important factor in corn production.

299. Smith, David B., et al. "Climatic Fluctuations and Agricultural Change in Southern and Central New England,

1765-1880." **Maine Historical Society Quarterly** 21 (Spring 1982): 179-200.

A study of the relationship between climate stress and New England agriculture. Based on the qualitative evidence found in diaries and records kept by amateur meteorologists that have been analyzed by quantitative techniques. Suggests that climate affected out-migration.

300. Teigen, Lloyd D. and Florence Singer. "Weather in U.S. Agriculture: Monthly Temperature and Precipitation by State and Farm Production Region, 1950-1990." United States Department of Agriculture, **Statistical Bulletin, No. 834**, 1992. 129 pp.

An excellent reference for atmospheric temperature and precipitation rates. Includes bibliography.

301. Tromp, S.W. "The Twentieth Anniversary of the International Society of Biometeorology: 1 January 1956-1 January 1976." **International Journal of Biometeorology** 20 (June 1976): 71-91.

Organized at the Biometeorological Research Center at Leiden, Netherlands, this group at first studied weather and environmental problems. Discusses the international cooperative work of the society.

302. Wang, J.Y, and Gerald L. Barger, ed. **Bibliography of Agricultural Meteorology**. Madison: University of Wisconsin Press, 1962. 673 pp.

    Dated but useful.

303. Whitnah, Donald R. **A History of the United States Weather Bureau**. Urbana: University of Illinois Press, 1960. 267 pp.

    A useful overview of the Weather Bureau's public service since 1870.

304. Whythe, Ian. "Human Response to Short- and Long-Term Climatic Fluctuations: The Example of Early Scotland." In **Consequences of Climatic Change**, edited by Catherine Delano Smith and Martin Parry, 17-29. Nottingham, England: Department of Geography, University of Nottingham, 1981. 143 pp.

    A study of the responses of rural communities in early-modern Scotland regarding short-term weather changes. Contends that these changes are important to understanding long-term agricultural adjustments to the weather.

CHAPTER IX

AGRICULTURAL CHEMISTRY

GENERAL

305. Anderson, O.E. **The Health of a Nation: Harvey Wiley and the Fight for Pure Food**. Chicago: University Chicago Press, 1958. 332 pp.

     An excellent history of Wiley's efforts to prevent the chemical adulteration of food. Good for the activities of the USDA's Bureau of Chemistry.

306. Aulie, Richard P. "Boussingault and the Nitrogen Cycle." **Proceedings of the American Philosophical Society** 114 (December 18, 1970): 435-479.

     Boussingault's studies of the chemical relationships between plants and animals brought agricultural science to the verge of modern microbiology. Between 1836 and 1876, his work on the importance of nitrogen in the life cycle became one of the greatest contributions to nineteenth-century science.

307. _____. "The Mineral Theory." **Agricultural History** 48 (July 1974): 369-382.

Discusses the work of Jean Baptiste Boussingault, Justice von Liebig, John Bennet Lawes and Joseph Henry Gilbert on the issue of whether mineral fertilizers could maintain soil fertility or whether they needed to be supplemented with nitrogenous sources.

308. Bassham, J.A., and M. Calvin. **Path of Carbon in Photosynthesis**. Elizabeth Cliffs, NJ: Prentice-Hall, 1957. 104 pp.

A study of the different aspects of carbon during photosynthesis. Includes historical background of the development of the carbon reduction cycle.

309. Bolton, Henry Carrington. **A Select Bibliography of Chemistry, 1492-1892**. Smithsonian Institution, Miscellaneous Collections, No. 851, 1893. 1,212 pp.

A dated but useful source for books and articles published in the United States and Europe before 1892. The majority of the publications are in English, French and German.

310. Borth, Christy. **Pioneers of Plenty: The Story of Chemurgy**. Indianapolis: Bobbs-Merrill Co., 1939. 303 pp.

A general history of chemurgy that provides an overview of the scientists who pioneered this field and the problems they confronted.

311. Browne, Charles A. "Bernhard Tollens (1841-1918) and Some American Students of His School of Agricultural Chemistry." **Journal of Chemical Education** 19 (June 1942): 253-259.

Tollens, director of the chemical laboratory of the Agricultural Institute at the University of Gottingen, attracted many students from the United States as well as from Great Britain, Russia, Japan and the Netherlands. The American attraction to Gottingen ended after Tollen's death.

312. _____. "European Laboratory Experiences of an Early American Agricultural Chemist--Dr. Evan Pugh (1828-1864)." **Journal of Chemical Education** 7 (March 1930): 499-517.

Pugh gained a scientific reputation for his experiments involving the fixation of atmospheric nitrogen by plants. This work is an abstract of Pugh's correspondence regarding his scientific observations in Europe.

313. _____, and Eva Armstrong. "History of Chemistry in America." **Journal of Chemical Education** 19 (August 1942): 379-81.

Traces the historical writing about chemistry from the 1874 work of Benjamin Sillman, Jr. through the studies of Henry Carrington Bolton, whom the authors consider the leading bibliographer of chemistry. Concludes with the influence of Edgar Fahs Smith, whom the authors contend became the most influential scientist in the history of chemistry in the United States.

314. Caldwell, George L. "The More Notable Events in the Progress of Agricultural Chemistry Since 1870." **Journal of the American Chemical Society** 14 (1892): 83-111.

An overview of the development of the American agricultural experiment stations, important chemical discoveries and leading scientists, such as Armsby, Van Bemmelen, Berthelot, Hilgard and Kellner.

315. Chamberlain, Joseph Scudder, ed. **Chemistry in Agriculture, A Cooperative Work Intended to Give Examples of the Contributions Made to Agriculture by Chemistry**. New York: The Chemical Foundation, 1926. 384 pp.

A general discussion of the many ways that chemistry affects agriculture, particularly in relation to crops, soils and diet. Calls soil, plants and animals "Nature's Great Triad," with chemical reactions in the soil sustaining the life of plants and animals.

316. Chittenden, Russell H. **The Development of Physiological Chemistry in the United States**. New York: Chemical Catalog Co., 1930. 427 pp.

A chronological discussion of the progress in physiological chemistry in the United States from 1880 to 1930.

Discusses the work of major scientists, but no critical analysis is applied.

317. Ferreira, Celia R.R.P. Tavares, Flavio Conde de Carvalho, and Antonio Jose Braga do Carmo. **Evolucao do Setor de Defensivos Agricolas no Brasil, 1964-1983**. Sao Paulo: Governo do Estado de Sao Paulo, Secretaria de Agricultura e Abastecimento, Instituto de Economia Agricola, 1986. 51 pp.

Describes the evolution of the agricultural chemical industry in Brazil.

318. Fruton, Joseph S. **Molecules and Life: Historical Essays on the Interplay of Chemistry and Biology**. New York: John Wiley & Sons, 1972. 579 pp.

A series of essays that trace the scientific activity designed to discover chemical explanations for biological phenomena. The author focuses on the period after 1800, because this experimental process became a major interest of leading scientists during the nineteenth century.

319. Fussell, George E. **Crop Nutrition: Science and Practice Before Liebig**. Lawrence, KS: Coronado Press, 1971. 232 pp.

Describes European agricultural practices in relation to soil improvement and chemical experimentation from antiquity to the mid-nineteenth century. A good introduction to the history of agricultural science before the formal development that field of inquiry.

320. Good, G.H. "On the Early History of Liebig's Laboratory." **Journal of Chemical Education** 13 (December 1936): 557-562.

Discusses the opening of laboratories prior to Liebig's, such as Thomas Thomson's about 1817 and Amos Eaton's in 1825. Includes a brief discussion of Liebig's work.

321. Gorlach, E., T. Mazur, and S. Moskal. "Chemia Rolna w 50-Leciu Polskiego Towarzystwa Gleboznawczego, 1937-1987." **Roczniki Gleboznawcze** 29 no. 2 (1989): 117-43.

A survey of the major developments in agricultural chemistry at the 50th anniversary of the Polish Society of Soil Science.

322. Guest, D., and B. Grant. "The Complex Action of Phosphonates as Antifungal Agents." **Biological Reviews of the Cambridge Philosophical Society** 66 no. 2 (1991): 159-87.

Surveys the history of phosphonates and their development as agricultural chemicals.

323. Hale, William J. **The Farm Chemurgic: Forward the Star of Destiny Lights Our Way**. Boston: Stratford Co., 1934. 201 pp.

Contends the industrial revolution brought the demise of the farm but that the rise of organic chemistry has given agriculture a rebirth. Considers the period from the late nineteenth to the early twentieth century and argues that agriculture will provide the basis for chemical research and testing in the twentieth century.

324. Holmes, Frederic L. "Elementary Analysis and the Origins of Physiological Chemistry." **ISIS** 54 (March 1963): 50-81.

Discusses the eighteenth- and nineteenth-century ideas, methods and experiments that led to the work of Liebig. Although the work of Liebig is emphasized, agricultural historians will profit from the chronological discussion of the work of others, such as Macquer, Berzellius, de Saussure and Boussingault.

325. Hurt, R. Douglas. "The Poison Squad." **Timeline** 2 (February/March 1985): 64-70.

Between 1902 and 1906, the Bureau of Chemistry in the USDA tested chemical food additives on groups of volunteers to help determine the effects of chemical preservatives on the public health. These tests helped achieve the Food and Drug Act of 1906.

326. Ihde, Aaron J. "Edmund Ruffin, Soil Chemist of the Old South." **Journal of Chemical Education** 29 (August 1952): 407-414.

Surveys Ruffin's experiments with lime or marl and manure to restore exhausted soil. Although his ideas were not always original, Ruffin publicized the benefits of correcting the acidity of the soil and encouraged the usefulness of "book farming."

327. Keuchel, Edward. "Chemicals and Meat: The Embalmed Beef Scandal of the Spanish American War." **Bulletin of the History of Medicine** 48 (Summer 1974): 249-264.

An excellent study of the canned meat controversy during the war along with a discussion of the meat packing industry and the refrigeration of beef.

328. Knight, D. "Agriculture and Chemistry in Britain Around 1800." **Annals of Science** [Great Britain] 33 no. 2 (1976): 187-196.

During the late eighteenth and early nineteenth centuries, scientists such as Cochrane and Davy urged farmers to apply chemistry to their farming practices rather than to rely on guess work. Davy particularly urged soil analysis by farmers.

329. Liebig, Justus von. **Chemistry in Its Applications to Agriculture and Physiology**. 2nd ed. London, 1842. 409 pp.

An essential study for historians of agricultural science. Liebig examines the supply of nutrients to plants and the changes that these chemicals undergo in the living organism. He also examines the changes that organic substances undergo in their conversion to inorganic compounds as well as the causes that determine these changes.

330. McMillen, Wheeler. **New Riches From the Soil: The Progress of Chemurgy**. New York: D. Van Nostrand Co., 1946. 397 pp.

A series of loosely organized narratives that report achievements in the field of chemurgy that brought commercial success. History of the concept of chemurgy and its significance.

331. Marco, Gino J., Robert M. Hollingworth, and Jack R. Plimmer, ed. **Regulation of Agrochemicals: A Driving Force in their Evolution**. Washington, D.C.: American Chemical Society, 1991. 188 pp.

A collection of symposium papers honoring the twentieth anniversary of the Division of Agrochemicals of the American Chemical Society. Good overview of law and legislation regulating pesticides in the United States.

332. Morrell, J.B. "The Chemist Breeders: The Research Schools of Liebig and Thomas Thomson." **Ambix** 19 (March 1972): 1-46.

A discussion of research schools based on the laboratory models of Liebig and Thomson. Includes information on financial support, publications and political power.

333. Moulton, Forest Ray. **Liebig and After Liebig: A Century of Progress in Agricultural Chemistry**. Washington, D.C.: American Association for the Advancement of Science, 1942. 111 pp.

Considers Liebig's influence on organic chemistry and his research on enzymes, nutrition, soils and fertilizers.

334. Multhauf, Robert. **The History of Chemical Technology: An Annotated Bibliography**. New York: Garland Publishing, 1984. 299 pp.

Includes introductory references for the history of agricultural chemistry and fertilizers. International in scope and includes author and subject index.

335. Munday, E. Patrick, III. "Sturm und Dung: (1803-1873): Justus von Liebig and the Chemistry of Agriculture." Ph.D. diss., Cornell University, 1990. 333 pp.

Discusses the relationship between Liebig's development of organic chemistry and the chemistry of agriculture. Emphasizes Liebig's work in plant physiology within the religious, social and political context of his time.

336. Phipps, T.T. "Commercial Agriculture and the Environment: An Evolutionary Perspective." **Northeastern Journal of Agricultural and Resource Economics** 20 (October 1991): 143-50.

A brief look at ground and surface water pollution from agricultural chemicals, including pesticides.

337. Rodricks, J.V. "FDA's Ban of the Use of DES in Meat Production: A Case Study." **Agriculture and Human Values** 3 (Winter/Spring 1986): 10-25.

Discusses the FDA ban in 1979 on dietylstibestrol, a drug that promotes growth in livestock. Describes the scientific discoveries that suggested DES was carcinogenic after the Food and Drug Administration approved the drug in 1962.

338. Rossiter, Margaret W. "The Charles F. Chandler Collection." **Technology and Culture** 18 (April 1977): 222-30.

A description of the Chandler Collection at Butler Library, Columbia University. Chandler's papers provide information on the professionalization of applied chemistry in the United States after 1860, especially for problems of adulteration and pollution.

339. _____. **The Emergence of Agricultural Science, Justus Liebig and the Americans, 1840-1880**. New Haven: Yale University Press, 1975. 275 pp.

Rossiter discusses the contributions of E.N. Horsford, J.P. Norton, S.W. Johnson and others as well as Justus von Liebig to the evolution of American agricultural science. America was quick to adopt Liebig's ideas on soil fertility but slow to establish institutions to support agricultural science, such as the experiment stations. Includes an appendix of Liebig's foreign students.

340. Simmons, I.G. "The Environmental Impact of Pre-Industrial Agriculture." **Outlook on Agriculture** 17 no. 3 (1988): 90-96.

Discusses the significance of agricultural chemicals to pastoral farming systems.

341. Skolnik, Herman, and Kenneth M. Reese. **A Century of Chemistry: The Role of Chemists and the American Chemical Society**. Washington, D.C.: American Chemical Society, 1976. 468 pp.

A history of the American Chemical Society. Includes the influence of governmental and public affairs as well as organizational structures and membership listings.

342. Snelders, H.A.M. "James F.W. Johnston's Influence on Agricultural Chemistry in the Netherlands." **Annals of Science** [Great Britain] 38 no. 5 (1981): 571-84.

Justus von Liebig's ideas on agricultural chemistry were introduced to the Dutch through translations of

Scotsman James Johnston's publications, during the mid-nineteenth century.

343. Summons, Terry G. "Animal Feed Additives, 1940-1966." **Agricultural History** 42 (October 1968): 305-313.

Discusses the reasons for the rise in meat production: improved technology; advances in veterinary medicine; improved nutrition; and the capital and skill to apply these scientific advancements. Stresses the scientific contributions to nutrition as the most important reason for the great increase in meat production.

344. Teich, M. "On the Historical Foundations of Modern Biochemistry." **Clio Medica** 1 (November 1965): 41-57.

Traces the origin of biochemistry to the late nineteenth century, when it separated from the science of chemical physiology after 1880.

345. Thompson, Malcolm J. "Contributions of USDA Scientists to the Chemistry and Biochemistry of Steroids and Other Isopentenoids." In **Isopentenoids in Plants: Biochemistry and Function**, edited by W. David Nes, Glenn Fuller and Lee-Shin Tsai, 3-27. Beltsville, MD: Insect Physiology Lab, United States Department of Agriculture, 1984.

Mainly credits a laboratory or research office rather than individual scientists, although the references provide the most important names and publications of the scientists working with plants and insects in this field.

346. Wells, C.B., and J.A. Prescott. "Origins and Early Development of Soil Science in Australia." **Soils: An Australian Viewpoint** (1993): 3-12.

Discusses agricultural chemistry and soil fertility and composition as well as the status of agricultural science at the Australian universities.

347. Wiley, Harvey W. **An Autobiography**. Indianapolis: Bobbs-Merrill Co., 1930. 339 pp.

Wiley's own story about his life as an agricultural chemist. Excellent background for studying the chemical problems of agriculture during the late nineteenth century and the response of the United States Department of Agriculture.

348. _____. "The Relation of Chemistry to the Progress of Agriculture." In **Yearbook of the United States Department of Agriculture, 1899**, 201-258. Washington, D.C.: Government Printing Office, 1900.

A review of the achievements in agricultural chemistry in the United States during the nineteenth century. Emphasis is given to the contributions of Liebig, Gilbert and Boussingault.

349. Williams, Robert R. "Chemistry as a Supplement to Agriculture in Meeting World Food Needs." **American Scientist** 44 (July 1956): 317-327.

Discusses the ways that chemistry can improve world food production, such as the enrichment of flour and corn meal. Argues that chemistry in nutritional research can help eliminate diseases such as pellagra and beriberi.

350. Wright, David E. "Waste Not, Want Not: The Michigan Roots of the Farm Chemurgic Movement." **Michigan History** 73 (September/October 1989): 32-38.

In 1925, William J. Hale began the farm chemurgic movement. Hale stressed the importance of chemistry to the development of new agricultural products. His research provided the basis for future investigations that led to the production of gasohol, biodegradable plastics, plywood and various products with soybean and cotton bases.

351. Young, James Harvey. **Pure Food: Securing the Federal Food and Drugs Act of 1906**. Princeton: Princeton University Press, 1989. 312 pp.

An essential study for historians of agricultural science who are interested in chemistry and public policy.

FERTILIZERS

352. Angus, J.F., C.F. St. Groth, and R.C. Tasic. "Between-farm Variability in Yield Responses to Inputs of Fertilizers and Herbicides Applied to Rainfed Lowland Rice in the Philippines."

Agriculture, Ecosystems and Environment [Australia] 30 no. 3-4 (1990): 219-34.

Discusses experiments in the central Philippines province of Antique which produced variable results. Suggests implications for extension education.

353. Becker, Pierre. **Phosphates and Phosphoric Acid: Raw Materials, Technology and Economics of the Wet Process**. New York: M. Dekker, 1989. 740 pp.

A highly technical classic on the production of phosphates and phosphoric acid.

354. Beeson, K.C. "The Effect of Fertilizer on the Nutritive Quality of Crops and the Health of Animals and Men." **Plant Food Journal** 5 (October/December 1951): 6-11.

Describes the mineral deficiencies in soils that affect the nutrition of plants and animals. Shows how fertilizers and other chemical compounds can correct these abnormalities for the benefit of human health.

355. Blakey, Arch F. **The Florida Phosphate Industry**. Cambridge: Harvard University Press, 1973. 197 pp.

Essentially an economic history of the industry from the 1880s through the 1960s. Historians of agricultural science will still find this study useful to help them understand the extractive aspects of the fertilizer industry.

356. Brand, Charles J. "Some Fertilizer History Connected with World War I." **Agricultural History** 19 (April 1945): 104-113.

Primarily a review of the use and regulation of nitrate by the War Industries Board. Includes a brief discussion of nitrogen research during the war.

357. Britton, E., ed. **Agriculture in Britain: Changing Pressures and Policies**. Wye, Kent, U.K.: Wye College, London University, 1990. 215 pp.

Discusses the pressures in Britain to limit the use of fertilizers. Contains ten essays. Good introduction to contemporary problems in British agriculture.

358. Carrier, Lyman. **The Beginnings of Agriculture in America**. New York: McGraw-Hill, 1923. 323 pp.

A general discussion that includes farm implements and the use of manure, lime and oyster shells for fertilizer. Dated but beneficial, if used with care.

359. Clarke, Margaret Jackson. "The Federal Government and the Fixed Nitrogen Industry, 1915-1926." Ph.D. diss., Oregon State University, 1977. 224 pp.

During World War I, the federal government sponsored nitrate research for the manufacture of explosives. When the fighting ceased, the War Department consolidated this research activity with the creation of the Fixed Nitrogen Research Laboratory in Washington D.C. to aid production for commercial purposes, including agriculture.

360. Collier, G.A. "Seeking Food and Seeking Money: Changing Productive Relations in a Highland Mexican Community." **Discussion Paper, United Nations Research Institute for Social Development, No. 11**, 1990. 23 pp.

Discusses the transformation of agriculture among the Zinacanteco Indians in the Chiapas highlands of southeastern Mexico. Describes the transformation from the labor intensive production of maize to dependence on fertilizers and herbicides to maintain productivity. Those who can afford this chemical technology have formed a new elite in local society.

361. Dazhong, W., and D. Pimentel. "Seventeenth Century Organic Agriculture in China. I. Cropping Systems in Jiaxing Region." **Human Ecology** 14 (March 1986): 1-14.

Discusses organic farming and cropping systems as well as the history of organic fertilizers in this region. Notes crop yields.

362. Donald, M.B. "History of the Chile Nitrate Industry." **Annals of Science** [Great Britain] 1 (January 15, 1936): 29-47.

Discusses the history of the commercial exploitation of Chile's nitrate deposits from pre-contact to the present, including the methods used and its agricultural benefits in various countries.

363. Emanuelsson, U., and J. Moller. "Flooding in Scania: A Method to Overcome the Deficiency of Nutrients in Agriculture During the Nineteenth Century." **Agricultural History Review** [Great Britain] 38 pt. 2 (1990): 127-48.

Analyses flood irrigation and surveys the history of manures for fertilizer in this area of Sweden.

364. Farber, Edward. "History of Phosphorus." **United States National Museum Bulletin, No. 240**, 177-200. Washington, D.C.: Smithsonian Institution, 1965.

A brief history of the discovery and use of phosphorous, including for agricultural purposes.

365. Feheir, Gyorgy. "A Mutragyazasi Technologia Kialakul alsa es a Mutragyak Hasznalatanak Elterjed else Magyarorszagon Az I. Vilaghaboru Elott." **Agrartorteneti Szemle** [Hungary] 27 no. 3/4 (1985): 395-434.

This article deals with the technological transformation of fertilizers for use in Hungary prior to World War I. As early as 1886, the Hungarian government advocated the use of fertilizers. Adoption, however, was not wide-spread because many imported fertilizers were either inappropriate for the climate or too expensive.

366. Fussell, G.E. "The Early Days of Chemical Fertilizers." **Nature** 195 (August 25, 1962): 750-754.

Before 1830, English farmers only used organic fertilizers. By that date, however, nitrate of soda was being imported from Chile as well as Peruvian guano. During the early 1850s, they also used a superphosphate that Justus von Liebig had advocated.

367. **Glossaire sur la Fertilisation**. Paris: IFA, 1986. 71 pp.

A glossary of fertilizer terms in French, English, Spanish and German languages.

368. Haynes, Williams. **Background and Beginnings, 1609-1911**. Vol. 1 of **American Chemical Industry**. New York: D. Van Nostrand Co., 1954. 512 pp.

A broad discussion of the chemical industry in America until the early twentieth century. Includes the history of chemical pioneering, the growth of the industry and the development of heavy and synthetic chemicals, electrochemicals and various fertilizers. See the other books in this 6 volume study for various historical aspects of the industry to 1948.

369. Jones, Grinnell. "Nitrogen: Its Fixation, Its Uses in Peace and War." **Quarterly Journal of Economics** 34 (May 1920): 391-431.

Discusses the use of nitrogen in Norway, Sweden, Switzerland, Germany, Austria, Japan, Italy, France, the United Kingdom and the United States since World War I. Notes considerations for the manufacture of explosives and fertilizer.

370. Jordan, Weymouth T. "The Peruvian Guano Gospel in the Old South." **Agricultural History** 24 (October 1950): 211-221.

Surveys the use of Peruvian guano in the late antebellum period. Although some southern farmers were being exposed to Edmund Ruffin's work with manure and while a few farmers experimented with this new fertilizer, guano was still not used on a wide basis. Includes a brief discussion of guano usage in South America and Europe prior to its arrival in the United States in 1820.

371. Keleti, Cornelius, ed. **Nitric Acid and Fertilizer Nitrates**. Vol. 4 of **Fertilizer Science and Technology Series**. New York: Marcel Dekker, Inc., 1985. 378 pp.

Traces the technological process for the production of fertilizer nitrate. Describes the early use of natural sodium nitrate from Chile and the most current methods for the production of ammonium nitrate and calcium ammonium nitrates. Includes a brief note on potassium nitrate.

372. Longworth, J.W., ed. **China's Rural Development Miracle, With International Comparisons**. Brisbane: University of Queensland Press for the International

Association of Agricultural Economists, 1989. 457 pp.

Contains 50 papers, many by Chinese scholars on the history of agriculture, including the use of fertilizers, in the People's Republic.

373. Maksudov, A. "Change in Thickness of Agroirrigation Deposits in the Fergana Valley Under Human Influence." **Problems of Desert Development** no. 1 (1988): 44-52.

Discusses the history of irrigated agriculture and the use of fertilizers in the Turkmen area of the Soviet Union.

374. Marcus, Alan I. "Setting the Standard: Fertilizers, State Chemists, and Early National Commercial Regulation, 1880-1887." **Agricultural History** 61 (Winter 1987): 47-73.

Follows the regulation of fertilizers beginning with the Grafton Mineral Fertilizer Company during the 1870s. Discusses the establishment of state chemistry bureaus and uniform testing by the agricultural experiment stations to determine the content of chemical fertilizers to enable regulation. Shows the conflict between USDA, state and commercial chemists over the manufacture and sale of chemical fertilizers.

375. Matlon, P.J. "Improving Productivity in Sorghum and Pearl Millet in Semi-Arid Africa." **Food Research Institute Studies** 22 no. 1 (1990): 1-43.

Discusses the research to increase crop production and the reasons for lack of success. Concludes that greater emphasis needs to be placed on the use of fertilizer to increase production and stabilize the agricultural base.

376. Molenar, J.D. de. "The Impact of Agrohydrological Management on Water, Nutrients, and Fertilizers in the Environment of the Netherlands." In **Agroecology: Researching the Ecological Basis for Sustainable Agriculture**, edited by S.R. Gliessman, 275-304. Berlin: Springer Verlag GmbH & Co., 1990.

A historical review of water management in the Netherlands and its influence on soil fertility, the use of fertilizers and the quality of ground water.

377. Munasinghe, A.R. **A Brief History and the Current Status of Ammonia Based Fertilizer Industry in Sri Lanka**. United Nations Industrial Development Organization, 1982. 5 pp.

Production of ammonia and urea fertilizers in Sri Lanka are discussed. Primarily economic rather than scientific.

378. Nelson, L.B. **History of the U.S. Fertilizer Industry**. Muscle Shoals, Alabama: Tennessee Valley Authority, 1990. 537 pp.

A general history of fertilizer from 1,000 B.C. to A.D. 1980. Emphasizes the industry in the United States since

1840. A detailed reference for teachers and students.

379. Pittman, W.E. Jr. "Technology and Resource Availability in the Florida Phosphate Industry. **Environment and Ecology** 7 no. 1 (1989): 145-150.

Reviews the changes in the Florida phosphate industry during the last century. Economic and environmental issues of this industry are addressed.

380. Pratap, Narayan. "Fertiliser Scenario--Past, Present and Future." In **Soil Fertility and Fertilizer**. Vol. 3 of **Nutrient Management and Supply System for Sustaining Agriculture in the 1990s**, edited by Virendra Kumar, G.C. Shotriya, and S.V. Kaore, 29-39. New Delhi: Indian Farmers Fertiliser Cooperative, 1990.

Describes the development of the fertilizer industry in India.

381. Quin, B.F. "The Quality of New Zealand Superphosphate." **New Zealand Agricultural Science** 16 no. 2 (1982): 93-100.

Discusses the recent history of the agricultural usage of superphosphates in New Zealand. Evaluates the problems of using this fertilizer from deposits on Christmas Island and the decreases in yield that have followed since the 1960s.

382. Randhawa, M.S. and I.P. Abrol. "Sustaining Agriculture: The Indian Scene." In **Sustainable Agricultural Systems**, edited by C.A. Edwards, et al., 438-50. Ankeny, IA: Soil and Water Conservation Society, 1990.

    Discusses the use of fertilizer to increase production of wheat and rice. Capital investments, however, often have outpaced increases in productivity.

383. Ruffin, Edmund. **An Essay on Calcareous Manures**. Edited by J.C. Sitterson. Cambridge: Harvard University Press, 1961. 199 pp.

    With the original publication of this book in 1832, Ruffin initiated an era of agricultural reform in the South. Ruffin conducted practical experiments with marl to improve soil fertility, and he published the results. Although Ruffin was not a chemist or scientist, his work helped farmers learn to improve soil fertility by the 1850s.

384. Schreiner, Oswald. "Early Fertilizer Work in the United States." **Soil Science** 40 (June-December 1935): 39-47.

    Discusses the early experiments during the eighteenth and nineteenth centuries with gypsum, guano, marl and manure. Brief notes on the contributions of C.T. Jackson at Yale, G.H. Cook at Rutgers University and others.

385. Sheridan, Richard C. "Chemical Fertilizers in Southern Agriculture." **Agricultural History** 53 (January 1979): 308-318.

Reviews the use of nitrogen, phosphorus and potassium fertilizers, and discusses the production of chemical fertilizers in Baltimore during the nineteenth century. Notes the activities of the Tennessee Valley Authority for developing new fertilizers.

386. Slack, Archie Vivian, and George Russell James, ed. **Ammonia**. Vol. 1 of **Fertilizer Science and Technology**. 4 Vols. New York: M. Dekker, 1973-1979.

Ammonia is one of the most important basic chemicals, particularly for fertilizer. This study provides both historical and highly technical information about production.

387. Stephens, Lester D. "Farish Furman's Formula: Scientific Farming and the 'New South.'" **Agricultural History** 50 (April 1976): 377-390.

During the late nineteenth century, this well-educated, Georgia-born farmer studied the use of fertilizer for the improvement of cotton production. He experimented with phosphoric acid, potash, lime, magnesium and nitrogen.

388. **Superphosphate--Its History, Chemistry and Manufacture**. United States Department of Agriculture, Agricultural Research Service and Tennessee Valley Authority. Washington, D.C.: Government Printing Office, 1964. 349 pp.

A critical review of publications about the industrial use of raw materials in the manufacture of superphosphate. It particularly shows the interrelationship of pure and applied research. Written for research and technical personnel, including agriculturists.

389. Tarr, Joel A. "From City to Farm: Urban Wastes and the American Farmer." **Agricultural History** 49 (October 1975): 598-612.

Discusses the use of human wastes for fertilizer in the United States and other countries. Few records exist, but the practice antedated the development of urban sewage systems.

390. Tie, Y.L., T.T. Ng, and C.C. Chai. "Hill Padi-Based Cropping System in Sarawak, Malaysia." In **Nutrient Management for Food Crop Production in Tropical Farming Systems**, edited by H. van der Heide, 301-12. Haren, Netherlands: Institute for Soil Fertility, 1989.

Describes the rice farming system in Sarawak and the use of fertilizers distributed by the broadcast method.

391. Tinsdale, Samuel L., and Werner L. Nelson. **Soil Fertility and Fertilizers**. 2nd ed. New York: Macmillan Co., 1966. 694 pp.

Reviews the basic principles of soil fertility and fertilizer. Assumes the reader has a solid understanding of agronomy.

392. Tsao, Lung-kung. **Fei Liao Shih Hua**. Beijing: Nung Yeh Ch Hsin Hau Shu Tien Pei-ching Fa Hsing So Fa Hsing, 1984. 74 pp.

See for the history of fertilizer in China. Includes bibliography of Chinese sources.

393. Turner, N.L. "Soil Fertility Studies Prior to 1895: January 1979-April 1990." **Quick Bibliography Series**. Beltsville, MD: National Agricultural Library, 1990. 22 pp.

See for sources on the history of fertilizer, distributors and soil fertility.

394. Uexkull, H.R. von. "The Fertilizer Situation in Asia and Its Effects on Small-Scale Farms." **Transfer of Technology to Small-Scale Farms**, 11-19. Taipei: Food and Fertilizer Centre for the Asian and Pacific Region, 1989.

Discusses fertilizer production and use in Asia and its effects on small-scale farmers. Evaluates the reasons for different usage in Kampuchea, Laos, Myanmar, Thailand, India, Vietnam, Bangladesh, Taiwan, Japan, Malaysia, the People's Republic of China, the Philippines and the Republic of Korea, especially in relation to economics and geographical isolation. Urges increased efficiency of application.

395. "War and the Fertilizer Industry, 1914-1939." **Fertilizer Review** 14 (September-October 1939): 4-5.

Prior to World War I, the United States was dependent on Germany for potash and Chile for nitrates. By 1939, however, the U.S. met its own nitrate needs from home production. Discusses ammonia production and the possible effects of the new war in Europe.

396. Welch, L.F. "Nitrogen: Ag's Old Standby." **Solutions Magazine** 30 (July/August 1986): 18-21.

A brief overview of the use of fertilizer in American agriculture. Perpetuates the myth about Squanto teaching the Pilgrims to plant corn with fish for fertilizer. Credits cheap nitrogen fertilizer with making American agriculture productive and profitable.

397. Wilson, Perry W. **The Biochemistry of Symbiotic Nitrogen Fixation**. Madison: University of Wisconsin Press, 1940. 302 pp.

Discusses the complex relationship between bacteria and leguminous plants. It is a sequel to **The Root Nodule Bacteria and Leguminous Plants** by E.B. Fred, I.L. Baldwin and Elizabeth McCory published in 1932.

398. Wines, Richard A. **Fertilizer in America: From Waste Recycling to Resource Exploitation**. Philadelphia: Temple University Press, 1985. 247 pp.

An excellent history of fertilizer use in the United States. Includes discussions of guano, superphosphates and recycling. Traces the development and maturation of the industry as a complex technological system. Contends that the nineteenth-century fertilizer industry operated as a monopoly similar to the twentieth-century OPEC oil cartel. Good for the discussion of technological and economic changes.

PESTICIDES

399. Boyce, A.M. "Historical Aspects of Insecticide Development." In **The Future for Insecticides: Needs and Prospects**, edited by Robert Metcalf and John McKelvey, Jr., 469-88. Vol. 6 of **Advances in Environmental Science and Technology**. New York: John Wiley & Sons, 1976.

    A historical survey of pesticide use from copper arsenate in 1867 to DDT after World War II. Discusses insecticide research in the public and private spheres.

400. Carson, Rachel. **Silent Spring**. Boston: Houghton Mifflin, 1962. 368 pp.

    An excellent study of chemical dangers to the environment. Carson began a debate that has not lessened over the use of chemicals in agriculture. Essential reading for anyone interested the history of pesticides and agriculture.

401. Chengxiang, Pan. "The Development of Integrated Pest Control in China." **Agricultural History** 62 (Winter 1988): 1-12.

    The People's Republic of China established a pest control program in 1952 that integrated traditional methods with western practices. The project continued into the 1970s with additional emphasis on limiting the chemical contamination of the environment.

402. Dahlsten, Donald L., and Richard Garcia, ed. **Eradication of Exotic Pests: Analysis With Case Histories**. New Haven: Yale University Press, 1989. 296 pp.

A collection of seventeen papers that generally view the eradication of pests in political and social terms, and as a scientific problem. Agricultural historians interested in the science, economics and politics of pest eradication will find this study useful.

403. Detroux, L. "De la Protection des Cultures a la Protection des Consommateurs." **Annales de Gembloux** 94 no. 1 (1988): 61-74.

Describes historical plant protection measures from antiquity to the present. Notes the dangers of pesticides, herbicides and fungicides and the research to determine the amounts of acceptable residues.

404. Downs, Eldon W., and George F. Lemmer. "Origins of Aerial Crop Dusting." **Agricultural History** 39 (July 1965): 123-135.

Discusses the dusting experiments of the United States Department of Agriculture and the U.S. Army Air Corps beginning in 1921. Includes material on recent aerial spraying.

405. Dunlap, Thomas R. **DDT: Scientists, Citizens and Public Policy**. Princeton: Princeton University Press, 1981. 318 pp.

An excellent discussion of the public debate over the ban of DDT. Although it emphasizes public policy, any scholar pursuing the history of agricultural science will find this study useful.

406. _____. "The Triumph of Chemical Pesticides in Insect Control, 1890-1920." **Environmental Review** 5 (1978): 38-47.

Argues that farmers adopted pesticides on a wide scale because chemicals were relatively cheap and easy to use in contrast to control methods that required changes in farming practices or increased education through extension services.

407. Endo, M. "Development into the Boundary Area of the Fertilizer and the Agricultural Chemical." **Proceedings of the Symposium on Fertilizer, Present and Future, September, 25-26, 1990, Tokyo**, 119-35. Tokyo: Japanese Society of Social Science & Plant Nutrition, 1991.

A history of the development and application of combination fertilizers and pesticides.

408. Fisher, S.W. "The Roots of Controversy: A Historical View of Pesticide Regulation." **American Nurseryman** 165 (January 1, 1987): 89-90, 92-94, 96-99.

An introductory review of legislation designed to help control insects. Begins with California's quarantine law in 1853 to control the coddling moth and concludes with the formation of the Environmental Protection Agency.

409. Forgash, A.J. "History, Evolution, and Consequences of Insecticide Resistance." **Pesticide Biochemistry and Physiology** 22 (October 1984): 178-186.

Scientific writings on the San Jose Scale and coddling moth date the pesticide resistance of anthropods to 1897. Notes the significance of resistance to world agriculture.

410. Frisbie, R.E., and P.L. Adkisson. "IPM: Definitions and Current Status in U.S. Agriculture." In **Biological Control in Agricultural IPM Systems**, edited by Marjorie A. Hoy and Donald C. Herzog, 41-51. Orlando, FL: Academic Press, 1985.

Provides background on Integrated Pest Management (IPM) during the 1970s and early 1980s. Describes various programs of the federal government for research and implementation of IPM; includes a discussion of economic and environmental incentives.

411. Goldman, Abe. "Tradition and Change In Postharvest Pest Management in Kenya." **Agriculture and Human Values** 8 (Winter/Spring 1991): 99-113.

The rapid loss of bushland, the growing emphasis on cash crops and an increasing population has caused traditional agricultural pest control techniques to be replaced by chemical pesticides. The application of science to agriculture in Kenya, however, has disrupted traditional farming and cultural practices.

412. Graham, Frank Jr. **Since Silent Spring**. Boston: Houghton Mifflin Co., 1970. 333 pp.

Discusses the work of Rachel Carson and her reaction to the environmental controversy that she created. Surveys the current status of pesticide regulation. Good introduction to federal policy, especial that established by the USDA.

413. Green, Maurice B. **Chemicals for Crop Improvement and Pest Management**. 3rd ed. New York: Pergamon Press, 1987. 370 pp.

Summarizes the use of pesticides in agriculture, particularly in relation to the United States and Great Britain. Includes a description of the types of pesticides and specific chemicals for particular applications and desired outcomes. Includes chapters on pest resistance and pesticide safety.

414. Hoy, M.A. "Biological Control of Arthropod Pests: Traditional and Emerging Technologies." **American Journal of Alternative Agriculture** 3 (Spring/Summer 1988): 63-68.

The biological control of anthropod pests remains an important agricultural tool. Genetic improvements of natural enemies via hybridization and recombinant DNA technology offer possibilities for great improvement in this field of pest control.

415. Ihde, Aaron J. "Pest and Disease Controls." In **Technology in Western Civilization**, edited by Melvin Kranzberg and Carroll Pursell, Jr., 369-385. Vol. 2. New York: Oxford University Press, 1967.

A general discussion of organic, inorganic and synthetic pest controls, such as Paris Green and DDT. Includes some discussion of the chemical history of weed control.

416. Jacobson, Martin. "Botanical Pesticides, Past, Present, and Future." In **Insecticides of Plant Origin,** edited by J.T. Arnason, B.J.R. Philogine and P. Morand, 1-10. American Chemical Society Symposium, Series No. 387. Washington, D.C.: American Chemical Society, 1989.

Discusses the most promising botanicals for the development of pesticides as well as their applicability. Contends botanical insecticides are safer than synthetics and reviews research in this area.

417. Knipling, Edward F. "The Eradication of the Screw-Worm Fly." **Scientific American** 203 (October 1960): 54-61.

Traces the scientific work that eliminated the screw-worm fly across the southern and southwestern United States.

Emphasizes the experiments of P.C. Bushland and D.E. Hoskins, USDA scientists at Kerrville, Texas, that involved the breeding and release of sterile flies.

418. Lang, T., and C. Clutterbuck. **P is for Pesticides**. London: Ebury Press in Association with The Pesticide Trust, 1991. 256 pp.

A consumer's guide to pesticides that the historian coming to this subject for the first time will find useful. Brief discussion of the history of specific pesticides, their presence in food, effects on health and alternatives. Includes a list of commonly used pesticides in the United Kingdom as well as the symptoms of pesticide poisoning.

419. Leon, Carlos, Laura D'Amato, and Maria E. Iturregui. "El Mercado de Plaguicidas en La Argentina." **Desarrollo Economic** [Argentina] 27 no. 105 (1987): 129-144.

The use of pesticides has become more prevalent in Argentina since the 1970's, and the authors note specific chemicals applied by farmers. This article not only looks at the pesticide industry in Argentina but it also considers the significance of Argentine pesticide exports to Latin America.

420. Lotti, M. "Production and Use of Pesticides." **NATO Agricultural Science Series, 11-17 Series H, Cell Biology** 13 (1987): 11-17.

Discusses the changes in the production and marketing of pesticides, herbicides and fungicides around the world since 1950. Includes the number of deaths from pesticide poisoning.

421. Maddy, Keith T., Susan Edmiston, and Donald Richmond. "Illness, Injuries, and Deaths from Pesticide Exposure in California 1949-1988." **Reviews of Environmental Contamination and Toxicology** 114 (1990): 57-123.

Since 1949, when California began keeping records about illnesses, injuries and deaths due to pesticide exposure, hundreds of new pesticides have been synthesized, tested and registered with the USDA and the California Department of Agriculture. As pesticide use increased, injuries and illnesses grew. Recent concerns involve chronic symptoms that cause reproductive problems and cancer.

422. Manners, Ian R. "The Persistent Problem of the Boll Weevil: Pest Control In Principle and In Practice." **Geographical Review** 69 (January 1979): 25-42.

Examines the ecological and historical background of boll weevil control. Contends that despite boll weevil resistance to chemical pesticides, this insect can be eliminated. Argues that chemical pesticides are ecologically safe.

423. Napompeth, B. "Use of Natural Enemies to Control Agricultural Pests in Thailand." Food and Fertilizer Technology Center [Bangkok], **Extension Bulletin No. 303**, 1990. 22 pp.

Discusses the history of biological pest control in Thailand. Arranged by crop headings. Lists the natural enemies of more than three dozen insects.

424. Ordish, George. **The Constant Pest: A Short History of Pests and Their Control**. New York: Charles Scribner's Sons, 1976. 240 pp.

A history of pest control from the Neolithic period to the late twentieth century. Includes a discussion of mechanical and chemical controls.

425. Packer, Kingsley. **Nanogen Index: A Dictionary of Pesticides and Chemical Pollutants**. Freedom, CA: Nanogens International, 1975. 256 pp.

A useful reference for agricultural chemicals and pesticides.

426. Perkins, J.H. "Naturally Occurring Pesticides and the Pesticide Crisis, 1945 to 1980." In **CRC Handbook of Natural Pesticides: Methods**, edited by N. Bhushan Mandava, 297-325. Vol. 5, **Naturally Occurring Pesticides**. Boca Raton, FL: CRC Press, 1985.

The pesticide crisis that developed after 1945 with the extensive use of synthetics resulted from technical and cultural phenomena. Technical problems included resistance, resurgence and secondary outbreaks, while cultural difficulties involved health and environmental dangers. Entomologists were the scientists most affected by this crisis. Although natural pesticides are less dangerous than synthetics, the public has been conditioned to believe that all pesticides are dangerous.

427. Pratchett, D. "DDT Concentrations in Cattle Grazing Irrigated Pastures in the Ord River Irrigation Area in Western Australia." **Australian Veterinary Journal** 67 no. 11 (1990): 423.

A note of the mobility and persistence of DDT in the food chain.

428. Primack, Joel, and Frank von Hipple. **Advice and Dissent: Scientists in the Political Arena.** New York: Basic Books, 1974. 299 pp.

In part, traces the battle between the USDA, Congress, the chemical industry and concerned citizens over the production and use of DDT. Although the federal government banned the use of DDT, the chemical industry contended that it did not cause cancer or birth defects. Argues that supporters of persistent pesticides still dominate federal policy.

429. Sawyer, Richard Clark. "To Make a Spotless Orange: Biological Control in California Oranges." Ph.D. diss., University of Wisconsin, 1990. 295 pp.

The state of California led the way for the biological control of pests. As early as 1889 ladybugs were imported to California from Australia to combat a scale insect that damaged citrus trees. Scientist Harry Scott Smith led the effort in California to use biological controls during the first half of the twentieth century.

430. Secoy, D.M., and Allan E. Smith. "Superstitions and Social Practices Against Agricultural Pests." **Environmental Review** 5 (1978): 2-18.

Reviews cultural influences on pest control worldwide from antiquity to the present. Emphasizes ancient cultural practices, such as prayer among the Romans.

431. Smith, Allan E., and D.M. Secoy. "Agricultural Pest Control in Colonial North America." **Journal of NAL Associates** 5 nos. 3-4 (1980): 71-75.

Describes the colonial methods for agricultural pest control, which included prayer, spells, incantations, cultivation, organic and inorganic substance applications and mechanical practices.

432. _____. "Salt as a Pesticide, Manure, and Seed Steep." **Agricultural History** 50 (July 1976): 506-516.

Discusses the use of salt in European agriculture from antiquity to the mid-nineteenth century, when farmers began to use other natural and man-made chemicals for pest control. British agricultural practices are emphasized.

433. Vietmeyer, Noel D. "Our 90-Year War With the Boll Weevil Isn't Over." **Smithsonian** 13 no. 5 (1982): 60-65, 67-68.

Traces the efforts to eradicate the boll weevil, which first appeared in Mexican cotton fields during the 1880s and arrived in Texas in the early 1890s. Although calcium arsenate helped control the boll weevil during the 1920s, the insect developed a resistance, and cotton farmers must still use chemical pesticides to protect their crops.

434. Whitaker, Adelynne H. "A History of Federal Pesticide Regulation in the United States to 1947." Ph.D. diss., Emory University, 1974. 473 pp.

Traces pesticide policy from the Progressive era with the Insecticide Act of 1910 to the federal Insecticide, Fungicide, and Rodenticide Act of 1947. An essential study for anyone investigating the history of pesticide control within the context of science, society and governmental policy.

435. _____. "Pesticide Use in Early Twentieth Century Animal Disease Control." **Agricultural History** 54 (January 1980): 71-81.

Discusses the efforts of the USDA's Bureau of Animal Industry from 1884 to 1910 to make pesticide manufacturers comply with minimum standards for quality control. The Insecticide and Fungicide Act of 1910 brought agricultural insecticides under the control of the federal government.

436. Whorton, James. **Before Silent Spring: Pesticides and Public Health in Pre-DDT America**. Princeton: Princeton University Press, 1975. 288 pp.

Traces the use of chemicals for pest control in agriculture before the introduction of DDT during the late 1940s. An excellent study that should be combined with the work of Carson and Dunlap.

437. Wilk, Valerie A., and Neville S. David. "Results of a Farmworker Survey on Pesticides: New Jersey 1983." **Migration World Magazine** 14 no. 3 (1986): 27-32.

Concludes that one-third of the migrant farmworkers in that state were sprayed with pesticides and that 70 percent of those workers reported health problems related to chemical exposure.

438. Worthing, Charles R., and S. Barrie Walker, ed. **The Pesticide Manual: A World Compendium**. 8th ed. Thornton Heath, Great Britain: The British Crop Protection Council, 1987. 1,081 pp.

Printed from the "Pesticide Database." It includes all known pesticides in use around the world. Notes development, toxicology and formulations. An excellent reference for agricultural historians.

HERBICIDES

439. Brian, R.C. "The History and Classification of Herbicides." In **Herbicides: Physiology, Biochemistry, Ecology**, edited by L.J. Audus, 1-54, Vol. 1. 2nd ed. New York: Academic Press, 1976.

A study of the development of organic and inorganic chemical herbicides, including a discussion of toxicity.

440. Buckholtz, K.P. "Weed Control--A Record of Achievement." **Weeds** 10 (July 1962): 167-70.

Outlines the importance of weed control as an agricultural discipline. Notes that more acres are chemically treated for weeds than for insects or plant diseases.

441. Cornstock, G. "Genetically Engineered Herbicide Resistance, Part One." **Journal of Agricultural Ethics** 2 no. 4 (1989): 263-306.

Outlines the moral principles concerning weed control and the breeding of herbicide-resistant crops. Discusses the difference between science and ethics and provides a brief history of weed control.

442. Crafts, Alden S. **Modern Weed Control**. Berkeley: University of California Press, 1975. 440 pp.

This is a second edition of a work first published in 1942. It updates weed control methods since the first publication. Includes an index of herbicides and a discussion of environmental problems.

443. Hill, G.D. "Impact of Weed Science and Agricultural Chemicals on Farm Productivity in the 1980's." **Weed Science** 30 no. 4 (1982): 426-29.

Traces the technological innovations in weed science during the late twentieth century, especially in relation to the emergence of the agrochemical industry. Argues that increased costs, negative publicity and governmental regulations are hindering the development of new herbicides for agriculture.

444. Johnson, George Fiske. "The Early History of Copper Fungicides." **Agricultural History** 9 (April 1935): 67-79.

Discusses the use of copper by the French as a fungicide during the early nineteenth century. Charles Morren is profiled as an advocate of copper sulfite to combat potato blight in 1845. Scientists and farmers used this chemical in the United States during the late 1850s or early 1860s.

445. Lilienfeld, D.E., and M.A. Gallo. "2,4-D, 2,4,5-T and 2,3,7,8-TCDD: An Overview." **Epidemilogical Reviews** 11 (1989): 28-58.

Surveys the history of three major herbicides and reviews data about cancer caused by exposure.

446. Lockhart, J.A.R., A. Samuel, and M.P. Greaves. "The Evolution of Weed Control in British Agriculture." In **Weed Control Handbook: Principles**, edited by R.J. Hance and K. Holly, 43-74. Oxford: Blackwell Scientific Publications, 1990.

Describes the history of weed control in Great Britain from cultivation to the application of herbicides about 1941. Some discussion of biological controls.

447. Parker, C. "The Role of Weed Science in Developing Countries." **Weed Science** 20 no. 5 (1972): 408-13.

A study of herbicide usage during the 1960s. Discusses the social impact of herbicides on the agricultural labor supply worldwide, especially in relation to peasant-based farming.

448. Peterson, Gale E. "The Discovery and Development of 2,4-D." **Agricultural History** 41 (July 1967): 243-53.

In April, 1942, P.W. Zimmerman and A.E. Hitchcock, scientists at the Boyce Thompson Institute, first reported the use of 2,4-D as an herbicide. The U.S. Department of Agriculture also began experiments with this chemical as an herbicide and, by 1946, scientists had proven its effectiveness.

449. Robinson, D.W. "The Impact of Herbicides on Crop Production. In **Opportunities for Increasing Crop Yields**, edited by R.G. Hurd, P.V. Biscoe, and D.W. Robinson, 297-312. London: Pitman, 1980.

A review of chemical weed control in early European history, including the effects of herbicides on soils and vegetable and fruit crops as well as the environment. Notes possibilities for future developments in weed control technology.

450. Whiteside, Thomas. **The Pendulum and the Toxic Cloud: The Course of Dioxin Contamination**. New Haven: Yale University Press, 1979. 205 pp.

Reviews the disaster at Seveso, Italy, when a 2,4,5-T manufacturing plant exploded. Includes the views of several scientists about the clean up of the area and discusses the possible long-term effects of this agricultural chemical on the local population. Directed at the popular audience. Most of this material previously appeared in **The New Yorker**.

CHAPTER X

PLANT SCIENCE

GENERAL

451. Arents, George. **Tobacco: Its History Illustrated by the Books, Manuscripts and Engravings in the Library of George Arents, Jr., Together With an Introductory Essay, a Glossary and Bibliographic Notes by Jerome E. Brooks**. 5 Vols. New York: Rosenbach Co., 1937-1952.

    An extensive bibliography on tobacco from 1507 to 1942. Volume 1: covers 1507 to 1615; Volume 2: 1615-1698; Volume 3: 1698-1783; Volume 4: 1784-1942; Volume 5: Index. Includes annotations; organized chronologically by publication date. Index contains names and subjects.

452. Agrawal, P.K. "Seed Industry in India: History, Policies and Perspectives." **Research Report, Instituut voor Ontwikkelingsvraagstukken, Katholieke**

Hogeschool Tilburg, No. 33, 1988. 56 pp.

A survey of India's seed Industry since the 1950s by the Indian Agricultural Research Institute. Emphasizes the influence of the National Seeds Corporation, the Green Revolution, the High Yielding Varieties Programme and national legislation. Makes recommendations for policy improvement.

453. Arrington, Leonard J. "Science, Government and Enterprise in Economic Development: The Western Beet Sugar Industry." **Agricultural History** 41 (January 1967): 1-17.

Surveys the growth of the sugar beet industry from 1888 to 1913. Economic in emphasis but some discussion of science and technology is included.

454. Aulie, Richard Paul. "Boussingault and the Nitrogen Cycle." Ph.D. diss., Yale University, 1968. 398 pp.

Jean Baptiste Boussingault's (1802-1887) research on the sources of plant nitrogen were the foundations for the modern understanding of the nitrogen cycle. This dissertation covers his long research career from the early 1820s until the 1870s.

455. Bender, George A. "Nineteenth Century Pharmaceutical Research into Plants of the Southwestern Deserts." **Pharmacy in History** 22 no. 2 (1980): 49-59.

A study of the research conducted by Parke, Davis and Company to develop medicinal products from the loco weed, California poppy, desert potato and manzanita and other plants in the American southwest, during the last two decades of the nineteenth century.

456. Bogue, Allan G. "Changes in Mechanical and Plant Technology: The Corn Belt, 1910-1940." **Journal of Economic History** 43 no. 1 (March 1983): 1-25.

Discusses the importance of the transition from horses to tractors, the adoption of hybrid varieties and the development of mechanical picker-huskers in the Midwest of the United States.

457. Bohm, Wolfgang. "Carl Sprengel als Wegbereiter der Pflanzenbauwissenschaft." **Zeitschrift fur Agrargeschichte und Agrarzociologie** [Federal Republic of Germany] 35 no. 2 (1987): 113-19.

Honors Carl Sprengel as a leading agricultural chemist during the early nineteenth century in Germany, particularly in relation to the science of plant nutrition.

458. Clark, Paul F. **Pioneer Microbiologists of America**. Madison: University of Wisconsin Press, 1961. 369 pp.

Discusses the early history of American microbiology via brief biographical sketches. Notes the importance of the work conducted at the agricultural experiment stations and the land grant colleges in relation to using bacteria to enrich top soil and to improve the dairy industry.

459. Coulter, J.M. "Development of Botany in the United States." **Proceedings of the American Philosophical Society** 66 (1927): 309-318.

A brief outline of the development of botany in the United States to the early 1920s. Useful for a quick overview.

460. Dryer, Peter. **A Garden Touched with Genius: The Life of Luther Burbank**. Berkeley: University of California Press, 1985. 293 pp.

A critical biography of Burbank's life and career. A good introduction to his contributions to agricultural science.

461. Ewan, Joseph A., ed. **A Short History of Botany in the United States**. New York: Hafner Publishing Co., 1969. 174 pp.

Includes fourteen essays on the development of botany to the 1960s. Historians of agricultural science will find the chapters on plant geography, ecology, and genetics to be particularly helpful.

462. Glass, Bentley. "The Strange Encounter of Luther Burbank and George Harrison Shull." **Proceedings of the American Philosophical Society** 124 no. 2 (1980): 133-53.

Because scientists questioned Burbank's methodology, George Harrison Shull was asked by the Carnegie Institution of Washington, D.C. to review his work. This study analyzes Shull's evaluation. Shull essentially concluded that while Burbank was honest and sincere in his methods, he was not always accurate or scientific in his work.

463. Groosman, R., A. Linnemann, and H. Wierema. "Seed Industry in Kenya." In **Seed Industry Development in a North/South Perspective**, 39-61. Wageningen, Netherlands: Center for Agricultural Publishing and Documentation, 1991.

Surveys the agricultural economy of Kenya in relation to the development of new seed varieties and government policy affecting plant breeders and seed certification. Despite a shortage of scientists, considerable progress has been made regarding maize, wheat, barley, rice and sunflower varieties. Urges the conservation of germplasm and the development of drought-resistant varieties.

464. Hampton, J.G., and D.J. Scott. "New Zealand Seed Certification." **Plant Varieties and Seeds** 3 no. 3 (1990): 173-80.

Traces the history of seed certification in New Zealand since 1915.

465. Hendricks, Sterling B. "The Passing Scene." **Annual Review of Plant Physiology** 21 (1970): 1-10.

A discussion of the changes in plant science during the early twentieth century, a transition period between classical study and modern science.

466. Johnson, Dale E. "Literature on the History of Botany and Botanic Gardens, 1730-1840: A Bibliography." **Huntia** 6 (September 1985): 3-121.

This issue is devoted to bibliography on the study of plants gathered for the Hunt Institute's master file. It accounts for virtually all botanical literature published from 1730 to 1840. Encyclopedia and dictionary entries have been excluded as well as biographies and treatises on plants.

467. Johnson, F. Roy. **The Peanut Story**. Murfreesboro, NC: Johnson Publishing Co., 1964. 192 pp.

A general history of the peanut from its origin in South America to the adoption of this crop worldwide. Economic emphasis but some discussion of technology.

468. Jones, H.A., and A.E. Clarke. "The Story of Hybrid Onions." In **Yearbook of Agriculture, 1943-1947**, 320-26. Washington, D.C.: United States Department of Agriculture, 1947.

Discusses the methods for producing hybrid onion seed from the 1920s to the 1940s, especially for developing the variety known as Italian Red. Includes some discussion of disease.

469. Krikorian, A.D. "Excerpts from the History of Plant Physiology and Development." In **Historical and Current Aspects of Plant Physiology: A Symposium Honoring F.C. Steward**, edited by P.J. Davies, 9-97. Ithaca: New York State College of Agriculture and Life Sciences, 1975.

A detailed and broad perspective on the foundation of plant physiology resulting from the adoption of chemistry by botanists during the nineteenth century, particularly those working in Germany. Includes a discussion of the early agricultural chemists who worked with nitrogen metabolism, especially John Kjeldahl.

470. Mangelsdorf, Paul C. **Plants and Human Affairs**. Nieuwland Lectures, Vol. 5. South Bend, IN: Notre Dame: University of Notre Dame, 1952. 29 pp.

Intriguing discussion of philosophical thought about man's dependence on plants. Although not based on scientific inquiry, Mangelsdorf reflects on the importance of the domestication of corn, the world food problem and plants that have special economic importance, such as rubber and tobacco.

471. Marshall, Charles E. **Microbiology for Agricultural and Domestic Science Students**. Philadelphia: P. Blakiston's Son and Co., 1911. 724 pp.

A textbook that is divided into three parts: morphological and cultural methods; physiological functions; and applied science. Dated but useful for historical comparisons.

472. Meidner, H. "Three Hundred Years of Research into Stomata." In **Stomatal Function**, edited by G.D. Farquhar and I.R. Cowan, 7-27. Stanford: Stanford University Press, 1987.

A history of the study of pores in the epidermis of plants, particularly leaves and stems.

473. Miller, G.W., and J.C. Pushnik. "The History of Plant Nutrition." In **Foliar Feeding of Plants with Amino Acid Chelates**, edited by H.D. Ashmead and Park Ridge, 11-23. Park Ridge, NJ: Noyes Publications, 1988.

A chronological record of the major contributions to plant nutrition since the age of Aristotle.

474. Miller, Robert H. **Root Anatomy and Morphology: A Guide to the Literature**. Hamden, CT: Shoe String Press, 1974. 271 pp.

This bibliography provides a comprehensive listing of the literature of root anatomy and morphology published since 1900 as well as a considerable number of publications that appeared earlier. Good for the study of research trends.

475. Morton, A.G. **History of Botanical Science: An Account of the Development of Botany from Ancient Times to the Present Day**. London: Academic Press, 1981. 474 pp.

A comprehensive history of botanical science, including its importance to agriculture.

476. Overfield, Richard A. "Charles E. Bessey: The Impact of the 'New' Botany on American Agriculture, 1880-1890." **Technology and Culture** 16 no. 2 (April 1975): 162-181.

Bessey, a professor, editor and writer, campaigned for the professionalization of agricultural science through the study of basic botanical principles. He urged cooperation between pure and applied agricultural scientists.

477. Pieters, A.J. "Seed Selling, Seed Growing, and Seed Testing." In **Yearbook of Agriculture, 1899**, 549-74. Washington, D.C.: United States Department of Agriculture, 1899.

An early history of the seed business in relation to American horticulture since the colonial period. Includes information about dealers, raising seeds and testing. Good overview.

478. Pinkett, Harold T. "Records of the First Century of Interest of the United States Government in Plant Industries." **Agricultural History** 29 (January 1955): 38-45.

Essential for any historian investigating the work of the USDA in relation to crop importations and the control of disease and pests.

479. Reed, Howard S. **A Short History of the Plant Sciences**. Waltham, MA: Chronica Botanica, 1942. 323 pp.

A brief history for graduate students. Covers basic discoveries and experiments.

480. Rodgers, Andrew Denny, III. **American Botany, 1873-1892**. Princeton: Princeton University Press, 1944. 340 pp.

A study of North American botany in relation to scholars who were influenced by Asa Gray, particularly regarding systematic investigations. The works of Parry, Palmer, Pringle, Macoun, Rusby, Green and Britton are emphasized.

481. _____. **John Torrey, A Story of North American Botany**. Princeton: Princeton University Press, 1942. 352 pp.

Torrey became the driving force for the systematization of North American botany. He classified most of the botanical materials gathered by the plant explorers of his age. Torrey became the most prominent American botanist in the nineteenth century.

482. _____. **Liberty Hyde Bailey: A Story of American Plant Sciences**. New York: Hafner, 1965. 506 pp.

A facsimile edition of the author's 1949 biography of Bailey. Useful for any historian of American agricultural science.

483. Singleton, W. Ralph. "Agricultural Plants." **Agricultural History** 46 (January 1972): 71-79.

Surveys the cultivation and adaptation of plants for agriculture in the American colonies. Includes contributions from the Native Americans, Africans and Europeans. Some discussion of technology.

484. Steere, William Campbell. **Fifty Years of Botany**. New York: McGraw-Hill, 1958. 637 pp.

A broad survey of the most important developments in American botany by forty scholars from various fields of plant science.

485. Troyer, J.R. "Bertram Whittier Wells (1884-1978): A Study in the History of North American Plant Ecology." **American Journal of Botany** 73 (July 1986): 1058-78.

Reviews the work of Wells who pioneered in the ecological study of the vegetation of the southeastern United States. Wells was part of the first generation of plant ecologists who worked during the first half of the twentieth century. Includes a complete list of his publications in this field.

486. Warburton, C.W. "A Quarter Century of Progress in the Development of Plant Science." **Journal of the American Society of Agronomy** 25 (January 1933): 25-36.

A discussion of the major advances in plant science in relation to the work of agronomists in the United States from approximately 1908 to 1933.

487. Wiley, Harvey W. "Sugar Beet: Culture, Seed Development, Manufacture and Statistics." United States Department of Agriculture, **Farmers' Bulletin, No. 52**, 1899. 40 pp.

Wiley, an ardent advocate of sugar beet production discusses the various methods, conditions, varieties, soil composition and fertilizers for this crop. Includes illustrations and an explanation of sugar beet technology for planting and cultivating.

488. Wilkes, Garrison. "Plant Germplasm Resources: American Independence, Past and Future." **Environmental Review** 1 (1976): 2-13.

Discusses the food surplus and the possibilities for increased crop yields as well as the advances in American agriculture since the 1830s.

489. Zeven, A.C. **Dictionary of Cultivated Plants and their Regions of Diversity: Excluding Most Ornamentals, Forest Trees and Lower Plants**. Wageningen: Centre for Agricultural Publishing and Documentation, 1982. 263 pp.

A useful source.

GENETICS and BREEDING

490. Babcock, Ernest B., and Roy E. Clausen. **Genetics in Relation to Agriculture**. New York: McGraw-Hill, 1918. 675 pp.

A dated text that briefly surveys a host of topics such as the fundamentals of heredity and plant and animal breeding, includes some historical discussion. Useful for comparing to contemporary research and writing.

491. Brown, William L. "USDA Contributions to Progress in Plant Genetics." **Agricultural History** 64 (Spring 1990): 315-18.

Review of plant breeding work to improve crops and enable resistance to disease.

492. Burton, G.W. "Plant Breeding, 1910-1984." In **Gene Manipulation in Plant Improvement**, edited by P. Gustafson, 1-15. New York: Pleum Press, 1934.

Emphasizes the application of Mendel's work to plant breeding. Includes reference to the development of disease-resistant Sudan grass for agriculture.

493. Cocking, E.C. "Hybridization Past and Present." In **Crop Species**, edited by P.V. Ammirato, et al., 1-8. Vol. 3 of **Handbook of Plant Cell Culture**. New York: Macmillan Publishing Co., 1984.

Briefly reviews the work of Luther Burbank and places him in the context of both prior and future hybridization research.

494. Cook, Robert. "A Chronology of Genetics." In **Yearbook of Agriculture, 1937**, 1457-77. Washington, D.C.: United States Department of Agriculture, 1937.

An introduction to the main developments in agricultural genetics to 1937. Designed for the layreader but scholars will find it useful.

495. Coons, George H. "The Sugar Beet: Product of Science (1837-1949)." **Scientific Monthly** 68 (March 1949): 149-164.

Traces the creation and breeding of sugar beets, with emphasis on techniques, especially breeding for disease resistance. Notes the development of refineries during the 1870s for sugar beets and surveys the production and use of this crop.

496. Corcos, A.F., F.V. Monaghan, and G. Mendel. Mendel's Work and Its Rediscovery: A New Perspective." **Critical Reviews in Plant Sciences** 9 no. 3 (1990): 197-212.

Examines the interpretations of Mendel's work since 1900. Contends that Mendel made conclusions about heredity based on empirical evidence rather than experimentation. Argues that geneticists during the twentieth century were responsible for the discovery of the so-called Mendel's laws.

497. Davis, Bernard D. **The Genetic Revolution: Scientific Prospects and Public Perceptions**. Baltimore: Johns Hopkins University Press, 1991. 295 pp.

A wide-ranging study of the applications of molecular genetics in biotechnology. Includes reflections about the future effects of genetics on agriculture for crop and livestock production.

498. Dunn, L.C. **A Short History of Genetics: The Development of Some of the Main Lines of Thought, 1864-1939**. Ames: Iowa State University Press, 1991. 261 pp.

A good historical review. This is a reprint of his study originally published by McGraw-Hill in 1965.

499. _____, ed. **Genetics in the 20th Century**. New York: Macmillan Co., 1951. 634 pp.

Essays on the progress of genetics during the first half of the twentieth century. Good for comparison with the status of genetics during the late twentieth century.

500. Elliott, Timothy James. "Genetics and Strain Improvement in the Genus Agaricus." Ph.D. diss., 1986. 268 pp.

The mushroom **Agaricus bisporus (Lange) Imbach** has been a major crop since 1650. The author provides detailed information on breeding strategy for strain improvement based on resistance to the fungicides carboxin, benodanil and tridemorph.

501. Evans, A.M. "The Genetic Variability of Crop Plants." In **Conservation of Plant Genetic Resources**, edited by J.C. Hughes, 8-13. Ashton, United Kingdom, British Association for the Advancement of Science, 1978.

A brief review of the consequences of domestication and mutation as the source of genetic variation in crop plants.

502. Fehr, Walter R., and Henry H. Hadley, ed. **Hybridization of Crop Plants**. Madison: American Society of Agronomy and the Crop Science Society of America, 1980. 765 pp.

Discusses the principles and procedures used to obtain hybrid seed. Intended for students, teachers and scholars in plant science. Analyzes the hybridization of the major agricultural plants.

503. Franci, C.M. "Conservation of Plant Genetic Resources: A Continuing Saga." **Journal of the Australian Institute of Agricultural Science** 52 no. 1 (1986): 3-11.

A brief discussion of the work of W.J. Farrer in Australia and N.I. Vavilov in the Soviet Union. Includes an outline of the Australian contribution to plant genetics and reviews threatened genetic resources, genetic engineering and forage plant breeding.

504. Goa, Y.T. "Yellow-seeded Rape in China." **Cruciferae Newsletter** no. 13 (1988): 34-35.

Outline of the recent history of breeding yellow-seeded rape to improve the yield of oil.

505. Gowen, John W. **Heterosis**. Ames: Iowa State College Press, 1952. 552 pp.

A collection of essays about the research during the early twentieth century to explain and utilize the hybrid vigor of plants.

506. Hartl, D.L., and V. Orel. "What did Gregor Mendel Think He Discovered?" **Genetics** 131 (June 1992): 245-253.

Reviews Mendel's thought. See for an introduction to plant breeding and historical perspective based on genetic research today.

507. Hawkes, J.G. "N.I. Vavilov: The Man and His Work." **Biological Journal of the Linnean Society** 39 no. 1 (1990): 3-6.

Brief note on his work and life, especially in relation to genetics, plant breeding, agroecology and the preservation of germplasm.

508. Janick, J., ed. **Plant Breeding Reviews**. New York: Van Nostrand Reinhold Co., 1987. 446 pp.

Contains 10 reviews on breeding and genetics, including durum wheat, barley, maize, and blueberries, some of which are set in historical context.

509. Kamps, M. "Plant breeding and Seed Production of Agricultural Crops in the Netherlands." **Prophyta** 6 no. 8 (1989): 4-19.

Emphasizes the historical role of the Dutch government, the agricultural institutes and the seed industry. Notes the publication of the first list of recommended varieties in 1924 and the Plant Breeder's Decree in 1942. Discusses the current status and institutional responsibilities.

510. Keller, Evelyn Fox. **A Feeling for the Organism: The Life of Barbara McClintock**. San Francisco: W.H. Freeman, 1983. 235 pp.

A good introduction to the genetic work of McClintock for the non-biologist as well as the scientist. Places McClintock's work in scientific and historical context.

511. Kimmelman, Barbara Ann. "A Progressive Era Discipline: Genetics at American Agricultural Colleges and Experiment Stations, 1900-1920." Ph.D. diss., University of Pennsylvania, 1987. 456 pp.

A history of the early development of the genetics departments at Cornell University, the University of Wisconsin and the University of California. These case studies emphasize the research shift at each institution from applied to pure research. This change occurred, in part, because agricultural breeders accepted Mendelian science as a means to improve productivity.

512. Krivchenko, V.I. "The Pride of Soviet Science." **Selektsiya-i-Semenovodstvo** [Moscow] 6 (1987): 39-43.

Outlines the work of Nikolai I. Vavilov, the Soviet pioneer geneticist (1887-1943). Vavilov traveled extensively and collected cereal grain specimens prior to World War II. In 1924, he helped found the institution that bears his name, the All-Union Vavilov Institute of Plant Industry (VIR), located in St. Petersburg.

513. Knudson, M.K. "The Role of the Public Sector in Applied Breeding R&D: The Case of Wheat in the USA." **Food Policy** 15 (June 1990): 209-17.

Discusses the pros and cons of public-sponsored, that is, government, research. Uses semi-dwarf and hybrid wheats as an example of the appropriate public role. Contends that public-sponsored research is essential when private institutions fail to pursue work that does not have immediate economic value.

514. Kloppenburg, Jack Ralph Jr., ed. **Seed and Sovereignty: The Use and Control of Plant Genetic Resources**. Durham: Duke University Press, 1988. 369 pp.

Addresses the control problem regarding genetic resources. The contributors provide a variety of opinions relating to the debate over the control of plant germplasm for commercial profit.

515. Levina, E.S. "From the History of Soviet Genetics: N.I. Vavilov and G.D. Karpechenko." **Soviet Genetics** 23 (May 1988): 1412-25.

Discusses the work of two major Russian plant breeders and reprints some of their correspondence. This article has been translated from **Genetika** 23 no. 11 (1987): 2007-19.

516. Li, Z.S. "Impact of Economic Policy on the Development of Genetics in China." **Genome** 31 no. 2 (1989): 898-99.

Notes the effect of the cultural revolution on the development of genetics, especially in relation to rice, wheat, oilseed rape, rubber, maize, soybean and tobacco crops.

517. Ludmerer, Kenneth M. **Genetics and American Society: A Historical Appraisal**. Baltimore: Johns Hopkins University Press, 1972. 222 pp.

Discusses the social history of genetics from 1905 to 1960. Analyzes the impact of science on American society through the use of genetic theories to determine social policy and evaluates the repercussions of social and political events for the study of genetics. Especially concerned with the influence of theories on the social views of the public and their influence on geneticists.

518. Mangelsdorf, Paul C. "Genetics, Agriculture, and the World Food Problem." **Proceedings of the American Philosophical Society** 109 no. 4 (1965): 242-48.

Mangelsdorf analyzes the effects of genetic improvements on increased crop and animal production, especially in the United States. He argues that hybrids and improved breeding practices, in conjunction with population control, will solve the problems of food shortages worldwide.

519. Morton, A.G. **History of Botanical Science: An Account of the Development of Botany from Ancient Times to the Present Day**. London: Academic Press, 1981. 474 pp.

A history of botany designed for the general reader rather than the scholar. Traces the origin of botany to the sixth century B.C. and provides a chronological discussion into the twentieth century. A good introduction that scholars will also find useful.

520. Olby, R.C. **Origins of Mendelism**. Chicago: University of Chicago Press, 1985. 310 pp.

First published in 1966, this second edition provides considerable new material on research since the mid-1960s. The bibliography is particularly good for studies of species multiplication by hybridization, which the author contends provides the scientific context for Mendel's research.

521. _____. "The Origins of Molecular Genetics." **Journal of The History of Biology** 7 (Spring 1974): 93-100.

Surveys the standard history of molecular genetics in order to trace important conceptual changes. Notes that the question of the exact roles of DNA and protein in the gene was more uncertain during the 1930s and 1940s than it appears from the vantage point of today.

522. Orel, V. "Gregor Mendels Erkenntnisse im Lichte der Entwicklung der Naturwissenschaft und Landwirtschaftslehre in der Ersten Halfte des 19 Jahrhunderts." **Wissenschaftliche Zeitschrift der Wilhelm Pieck Universitat Rostock, Naturwissenschaftliche Reihe** 34 no. 2 (1985): 5-11.

A discussion of Mendel's perceptions on genetics in relation to agricultural science during the first half of the nineteenth century.

523. Richards, John, ed. **Recombinant DNA: Science, Ethics, and Politics**. New York: Academic Press, 1978. 368 pp.

A collection of essays that study the recombinant DNA controversy from the perspective of ethics and the decision-making process. Central to the debate is the need for the researcher and policy maker to evaluate the benefits and risks when there are unknown factors relating to the creation of new life forms.

524. Robbelen, S., R.K. Downey, and A. Ashri, ed. **Oil Crops of the World: Their Breeding and Utilization**. New York: McGraw-Hill, 1989. 553 pp.

A detailed study with some historical context. Eighteen crops are discussed.

525. Roberts, H.F. **Plant Hybridization Before Mendel**. Princeton: Princeton University Press, 1929. 374 pp.

A detailed analysis of plant hybridization before the discovery of Mendel's papers in 1900. Limited to the work that contributed to theoretical discoveries in plant fertilization and hybridization.

526. Rosenberg, Charles E. "Factors in the Development of Genetics in the United States: Some Suggestions." **Journal of the History of Medicine** 22 (January 1967): 27-46.

Discusses the American contribution to the development of this subfield of biology, during the early twentieth century. Emphasizes social effects in relation to institutional research. Contends that genetics emerged with the creation of modern America because of technological and economic advantages.

527. Rousset, M. "Les Connaissances Scientifiques et Techniques en Matiere de Selection Semenciere." **Actes et Communications, Economie et Sociologie Rurales, Institut National de la Recherche Agronomique** no. 4 (1989): 9-17.

Surveys the history of plant breeding in France, with a focus on wheat and maize. Contends yields increased with improved breeding and better cultural techniques.

528. Sears, E.R. "Genetics and Farming." In **Science in Farming: Yearbook of Agriculture, 1943-1947**, 245-255. Washington, D.C.: United States Department of Agriculture, 1947.

A popular look at the manner in which genetics has improved agricultural progress and productivity. Directed at the general reader rather than the scholar.

529. Sprague, G.F. "Breeding for Food, Feed, and Industrial Uses." In **Seeds: Yearbook of Agriculture**, 119-27. Washington, D.C.: United States Department of Agriculture, 1961.

Discusses the major improvements in corn, wheat, cotton and potatoes to meet specific needs since the eighteenth century.

530. Singleton, W. Ralph. "The Use of Radiation in Plant Breeding." In **Atomic Energy and Agriculture**, edited by C.L. Comar, 183-94. Washington, D.C.: American Association for the Advancement of Science, 1957.

Illustrates the direct application of ionizing radiation to the improvement of crops. Notes studies with barley and corn. Discusses future applications.

531. Srom, Frantisek. "Prukopnik Sovetske Genetiky. K 100. Vvroci Narozeni N.I. Vavilov." **Slovansky Prehled** [Czechoslovakia] 74 no. 4 (1988): 343-52.

In 1929, Nikolai I. Vavilov helped found the Academy of Sciences Institute of Genetics in the Soviet Union. Prior to World War II, he was a pioneer Soviet plant breeder who sought increased agricultural productivity. In 1943, he was executed for his bourgeois ideology.

532. Stubbe, Hans. **History of Genetics: From Prehistoric Times to the Rediscovery of Mendel's Laws**. Translated by Trevor R.W. Waters. Cambridge: M.I.T. Press, 1972. 356 pp.

A translation of **Kurze Geschichte der Genetik bis zur Weiderentdeckung der Vererbungsregeln Gregor Mendels**. A chronological study designed for the student and scholar as an introduction to the history of genetics.

533. Sturtevant, Alfred Henry. **A History of Genetics**. New York: Harper & Row, 1965. 165 pp.

Surveys Mendel's work and its influence to the mid-twentieth century.

534. Thomas, Dena Rae. "Plant Breeders and Plant Geneticists: Biographical Information." **Science & Technology Libraries** 11 (Winter 1991): 137-53.

Brief biographical essays on scientists such as Barbara McClintock and N.I.

Vavilov. Included are selected works by each scholar as well as selected studies about them.

535. Vavilov, Nikolai Ivanovich. **Epistolary Heritage, 1911-1928**. N.P.: Nauchnoe Nasledstvo, 1980. 428 pp.

A volume of the author's early correspondence. Emphasizes his contributions to the organization of biological and agricultural science in the Soviet Union. A useful study for anyone interested in the history of Soviet science.

536. _____. **The Origin, Variation, Immunity and Breeding of Cultivated Plants: Selected Writings**. Translated by K. Starr Chester. Waltham, MA: Chronica Botanica, 1951. 364 pp.

A dated but useful reference.

537. Warmbrodt, Robert D., and Lara Wiggert. **Biotechnology: Genetic Engineering for Crop Plant Improvement, January 1991-March 1992**. Quick Bibliography Series--92-48. Beltsville, MD: National Agricultural Library, 1992. 39 pp.

A useful guide to the recent literature.

538. Watson, James D., and John Tooze. **The DNA Story: A Documentary History of Gene Cloning**. San Francisco: W.H. Freeman and Co., 1981. 605 pp.

A collection of documents, articles, cartoons, letters and speeches covering the DNA and recombinant DNA controversy. Includes a section of definitions for the lay person as well as explanations of the scientific significance of recombinant DNA in regard to biology.

539. Webber, Herbert J, and Ernest A. Bessey. "Progress of Plant Breeding in the United States. In **Yearbook of Agriculture, 1899**, 465-90. Washington, D.C.: United States Department of Agriculture, 1899.

    A good review for historical context and comparison with current research activities.

540. White, G.A., H.L. Shands, and G.R Lovell. "History and Operation of the National Plant Germplasm System." **Plant Breeding Reviews** 7 (1989): 5-56.

    A study of the creation of the National Plant Germplasm System in the United States. Includes a discussion of plant introductions, collection and storage as well as matters of quarantine, seed exchange and general operations. This institution resulted from the southern corn blight epidemic in 1970 that showed the need for genetic diversity.

541. Williams, J.T. "Plant Genetic Resources: Some New Directions." **Advances in Agronomy** 45 (1991): 61-91.

    A short history of the efforts to preserve plant germplasm. Discusses the significance of this work for increasing and stabilizing production.

542. Yeatman, C.W., D. Kafton, and G. Wilkes, ed. **Plant Genetic Resources: A Conservation Imperative**. Boulder: Westview Press, 1984. 164 pp.

A collection of eleven essays, some of which discuss genetic conservation and the breeding of crops plants as well as the introduction of plants into Canada and the United States.

543. Zirkle, Conway. **The Beginnings of Plant Hybridization**. Philadelphia: University of Pennsylvania Press, 1935. 231 pp.

Discusses plant hybridization prior to 1761. The author acknowledges that some of these early efforts to develop hybrids are worthless, but they are useful for tracing the history of hybridization. Analyzes the work of Koelreuter, Linnaeus, Bartram and others.

544. _____. "Plant Hybridization and Plant Breeding in Eighteenth Century American Agriculture." **Agricultural History** 43 (January 1969): 25-38.

Although the word hybrid was first used about 1737, colonial farmers did not understand that concept. Still, Cotton Mather and others recognized the rudimentary principles of hybridization, even though they could not explain it.

PLANT INTRODUCTION

545. Carleton, Mark. "Russian Cereals Adapted for Cultivation in the United States." United States Department of Agriculture, Division of Botany, **Bulletin 23**, 1900. 42 pp.

Describes the twenty-three varieties of cereals that the author collected in Russia, including wheat, barley, rye, emmer, maize, millet, buckwheat and flat peas. Includes a useful comparison of Russian and American soils and climate in the Steppes and Great Plains as well as a discussion of Russian harvesting techniques.

546. Creech, John L. "The Greatest Service." In **After a Hundred Years: Yearbook of Agriculture, 1962**, 100-05. Washington, D.C.: United States Department of Agriculture, 1962.

Traces the work of Frank N. Meyer, who worked in China from 1905 to 1913. Meyer sought plants that would benefit American agriculture. Brief discussion of the technical aspects of plant collecting.

547. Cunningham, Isabel Shipley. "Research Reveals New Information About Frank N. Meyer, Agricultural Explorer in Asia." **Journal of NAL Associates** 5 no. 3/4 (1980): 79-83.

Brief sketch of the Dutch-born plant explorer who traveled in Asia for ten years during the early twentieth century for the USDA.

548. Dillman, A.C. "The Beginnings of Crested Wheatgrass in North America." **Journal of the American Society of Agronomy** 38 (March 1946): 237-50.

This hardy, drought-resistant and soil-holding Siberian grass was introduced to the United States in December 1906. Between 1908 and 1915, plantings were made at the Belle Fourche Station at Newell, South Dakota, and at the Northern Great Plains Field Station at Mandan, North Dakota. The early distribution of this grass seed in the Northern Great Plains came from these stations.

549. Ewan, J., and Nesta Ewan. "John Lyon, Nurseryman, and Plant Hunter, and His Journal, 1799-1814." **Transactions of the American Philosophical Society** 53 Part 2 (May 1963): 1-69.

A large part of this article is a reprint of Lyon's journal which shows his day-to-day activities as a plant explorer in the south and central Atlantic states. Includes a list of horticultural introductions and new species.

550. Fairchild, David. **Exploring for Plants**. New York: Macmillan Co., 1930. 591 pp.

A general history of the author's travels and work as a plant explorer. Written for the layman as well as the specialist. Dated but useful.

551. _____. "Systematic Plant Introduction, Its Purposes and Methods." United States Department of Agriculture, Division of Forestry, **Bulletin 21**, 1898. 24 pp.

Discusses recent progress in plant breeding and pathology and explains the purposes of plant introduction. Includes an explanation of the methods of plant collection, inspection, instruction and distribution.

552. _____. **The World Was My Garden: Travels of a Plant Explorer**. New York: Charles Scribner's Sons, 1938. 494 pp.

An autobiography through 1930. Discusses his extensive travels in Asia, Europe, Africa, North America and South America as a plant explorer.

553. Gilstrap, Marguerite. "The Greatest Service to Any Country." In **Seeds: Yearbook of Agriculture**, 18-27. Washington, D.C.: United States Department of Agriculture, 1961.

A general discussion of the search for germplasm to improve crop breeding. Highlights the work of John Bartram, Eliza Lucas, Thomas Jefferson, William Baldwin, Theodonck Bland and others. Notes the introduction of wheat, rice and soybeans from abroad.

554. Harlan, Jack R. "The Utility of Plant Exploration." **Journal of NAL Associates** 8 (January/ December 1983): 199-208.

Discusses plant introductions around the world and advocates an American program for exploration and the introduction of new plant varieties to diversify agriculture.

555. Healey, Ben J. **The Plant Hunters**. New York: Charles Scribner's Sons, 1975. 214 pp.

An illustrated history that includes a list of plant names. Surveys the work of plant hunters such as John Tradescant, George Egehard Rumpf and John Bartram. Notes the failures, friendships and jealousies of the plant collectors. A useful introduction.

556. Hodge, W.H. and C.O. Erlanson. "Federal Plant Introduction: A Review." **Economic Botany** 10 (October-December 1956): 299-334.

An excellent study of plant introductions in the United States since 1898, with emphasis on hard winter wheat and soybeans as successful and economically important introductions. See for the contributions of individual scientists.

557. Hodge, W.H., et al. "Federal Plant Introduction Gardens." **National Horticultural Magazine** 35 (April 1956): 86-106.

Describes the gardens used by USDA scientists to test new or little-known plants. General discussion of the investigations conducted at the plant introduction gardens at Coconut Grove, Florida, Chico, California, Glenn Dale, Maryland, and Savannah, Georgia.

558. Hyland, Howard L. "History of United States Plant Introduction." **Environmental Review** 4 (1977): 26-33.

Hyland discusses four phases of plant introduction: the work of plant scientists prior to formalization of a USDA program in 1898; explorations from 1898 to 1930; activities between 1930 and 1948; and the U.S. cooperative program from 1948 to 1975. A good, brief introduction.

559. Klose, Nelson. **America's Crop Heritage: The History of Foreign Plant Introduction by the Federal Government**. Ames: Iowa State College Press, 1950. 156 pp.

A survey of the Bureau of Plant Industry's work to introduce new plants for agricultural purposes. Designed for the general reader. Shows America's agricultural debt to foreign nations. A useful introduction.

560. _____. "Experiments in Tea Production in the United States, 1800-1912." **Agricultural History** 24 (July 1950): 156-161.

Traces the efforts to introduce tea to American agriculture from the early nineteenth century to 1912. Notes introductions at various locations.

561. Larsen, Esther L. "Peter Kalm's Short Account of the Natural Position, Use, and Care of Some Plants, of Which the Seeds Were Recently Brought Home from North America for the Service of Those

Who Take Pleasure in Experimenting with the Cultivation of the Same in Our Climate." **Agricultural History** 13 (January 1939): 33-64.

Essentially Kalm's pamphlet published in 1751. Kalm listed 126 American plants by the common name and gave a description of each to encourage their introduction to his native Sweden.

562. Li, H.L., and A.F. Stanley. "Russo-United States Botanical Exchange in 1841: Some Material Evidence." **Morris Arboretum Bulletin** 26 (December 1975): 51-56.

Examines the failed efforts of the director of the St. Petersburg Imperial Botanical Garden and the secretary of the American legation to introduce Russian seeds into the United States.

563. Mack, R.N. "Catalog of Woes: Some of our Most Troublesome Weeds were Dispersed Through the Mail." **Natural History** 3 (1990): 44-53.

Traces the beginnings of mass mailings of catalogs and seeds to the late 1840s. Farmers sought new plants for food and fiber, but poppies and hemp eventually caused more problems than economic benefits. Describes other plant introductions that caused difficulties, such as the barberry and goat grass.

564. Rasmussen, Wayne D. "Diplomats and Plant Collectors: The South American Commission, 1817-1818." **Agricultural History** 29 (January 1955): 22-31.

The commission served as a diplomatic and fact-finding body to keep the U.S. informed about revolutionary activities and to support the new governments, but it also provided the first botanical exploration in South America by a trained botanist, William Baldwin. Overall, the commission's horticultural explorations proved more beneficial to the United States than its political mission.

565. Ryerson, Knowles A. "History and Significance of the Foreign Plant Introduction Work of the United States Department of Agriculture." **Agricultural History** 7 (January 1933): 110-28.

Mentions early plant introductions during the colonial period, but traces serious work to 1836, when the Patent Office began collecting seeds for distribution. Discusses the post-Civil War work of William Saunders and James Wilson.

566. _____. "Plant Introductions." **Agricultural History** 50 (January 1976): 248-57.

A survey of plant introductions to the U.S. since the colonial period. Notes the significance of John Bartram and Henry C. Ellsworth as well as the importance of the Morrill and Hatch acts.

567. Stetson, Sarah P. "Traffic in Seeds and Plants from England's Colonies in America." **Agricultural History** 23 (January 1949): 45-56.

Traces the introduction of unusual British plants and seeds to the U.S. Emphasizes the work of botanist John Bartram, who, in 1728, planted the first botanical garden in America near Philadelphia.

568. Taylor, Mrs. H.J. "To Plant the Prairies and the Plains: The Life and Work of Niels Effesen Hansen." **South Dakota Historical Collection** 21 (1942): 185-290.

A brief discussion of Hansen's life. In 1934, he worked with Soviet scientists to develop a perennial wheat and improve Soviet agriculture. As a USDA scientist Hansen introduced the Manchu apricot and alfalfa as well as varieties of pears, plums, cherries, apples, melons and the tailless sheep to the prairie/plains region of the United States from Russia and China.

569. Wright, W.F. "The History of the Cowpea and Its Introduction into America." United States Department of Agriculture, Bureau of Plant Industry, **Bulletin 102** (1907): 43-59.

Traces the origin of the cowpea to Afghanistan and its introduction to the Western Hemisphere during the late seventeenth century. It reached the North American continent from the West Indies during the first half of the eighteenth century.

PLANT PATHOLOGY

570. Ainsworth, G.C. **Ainsworth & Bisby's Dictionary of Fungi**. 6th ed. Farnham Royal, England: Commonwealth Agricultural Bureaux for the Commonwealth Mycological Institute, 1971. 663 pp.

The standard dictionary for mycology. Includes material on lichens by P.W. James and D.L. Hawksworth.

571. _____. **Introduction to the History of Mycology**. Cambridge: Cambridge University Press, 1976. 359 pp.

The first one-volume history of mycology. Chapter topics include the origin of fungi, nutrition, genetics, distribution and classification of fungi. Includes a chronology and bibliography for the period 1491-1974. Useful for plant pathologists and historians of agricultural science.

572. _____. **Introduction to the History of Plant Pathology**. Cambridge: Cambridge University Press, 1981. 315 pp.

Discusses the history of plant pathology as an applied agricultural science for the diagnosis, treatment and prevention of disease.

573. Campbell, R. **Biological Control of Microbial Plant Pathogens**. Cambridge: Cambridge University Press, 1989. 218 pp.

An introduction to the history and practice of plant pathology and microbial ecology.

574. Carefoot, Garnet L., and Edgar Sprott. **Famine on the Wind: Man's Battle Against Plant Disease**. Chicago: Rand McNally, 1967. 231 pp.

A review of the most important fungi and viruses that have affected agriculture worldwide.

575. Chand, J.N. "Phytobacteriology in India--Past Present and Future Prospects." **Indian Phytopathology** 42 no. 1 (1989): 1-15.

A brief history of research on bacterial crop diseases in India.

576. Conners, I.L., ed. **Plant Pathology in Canada**. Winnipeg, Canada: Canadian Phytopathological Society, 1972. 251 pp.

A geographical history arranged by provinces. Emphasizes the people rather than their research or publications. Good biographical information.

577. Dunegan, John C., and S.P. Doolittle. "How Fungicides Have Been Developed." In **Plant Diseases: Yearbook of Agriculture, 1953**, 115-20. Washington, D.C.: United States Department of Agriculture, 1953.

Surveys the late nineteenth-century work of French scientists at the University of Bordeaux to develop a fungicide for the protection of fruits and vegetables from diseases. Includes a discussion of the compounds used since 1930.

578. Fawcett, Howard S. **Adventures in the Plant Disease World**. Berkeley: University of California Press, 1941. 34 pp.

A lecture presented to the Faculty Research Group at the University of California at Los Angles on March 27, 1940. Describes some of the plant diseases in the world, especially the potato blight fungus in Ireland during the 1840s. Includes historical material on fungus and bacterial diseases.

579. Fernandez-Valiela, M.V. "Algunas Consideraciones Sobre el Desarrollo de la Fitopatologia en la Argentina." **Fitopatologia Brasilera** 13 no 3. (1988): 177-79.

Reviews the development of plant pathology in Argentina since 1881.

580. Fuchs, W.H. "History of Physiological Plant Pathology." **Encyclopedia of Plant Physiology**. New Series 4 (1976): 743-59.

The history of plant pathology must include the study of physiology and biochemistry. Emphasizes the work of Duggar, who, as early as 1911, stressed the need to study physiological pathology and chemical pathology to understand plant diseases.

581. Galloway, Beverly T. "Plant Pathology: A Review of the Development of the Science in the United States." **Agricultural History** 2 (April 1928): 49-60.

Discusses the early work in the USDA, particularly for developing plant pathology as a scientific discipline.

582. _____. "Progress in the Treatment of Plant Diseases in the United States." In **Yearbook of Agriculture, 1899**, 191-200. Washington D.C.: United States Department of Agriculture, 1899.

Surveys the classification and the early theories that explained plant disease, such as the barberry bush as the cause of wheat rust. Traces the beginning of modern pathology to 1845.

583. Gudmestad, N.C. "A Historical Perspective to Pathogen Testing in Seed Potato Certification." **American Potato Journal** 68 no. 2 (1991): 99-102.

Historical note on potato seed certification around the world. Some discussion of disease and diagnosis.

584. Holmes, C.S., et al., ed. **Plant Pathology: Problems and Progress, 1908-1958**. Madison: University of Wisconsin Press, 1959. 588 pp.

A collection of essays on ten major subjects, including history, parasitism, genetics, fungicides, soil microbiology, nematology and diseases. Authors include scientists from American, English, Australian, Mexican, Dutch,

Canadian, Honduran, German and South African institutions. More scientific than historical in coverage but useful.

585. Jones, L.R. "Biographical Memoir of Erwin Frink Smith, 1854-1927." **National Academy of Sciences, Biographical Memoirs** 21 (1941): 1-71.

Summarized his research on peaches, potatoes, cabbages and diseases in plants from 1886 to 1922. Includes a list of his publications.

586. Horsfall, James G., and Ellis B. Cowling. **Plant Diseases: An Advanced Treatise**. 5 Vols. New York: Academic Press, 1977-1980.

A comprehensive study. See, Vol. 1: How Disease is Managed; Vol. 2: How Disease Develops in Populations; Vol. 3: How Plants Suffer from Disease; Vol. 4: How Pathogens Induce Disease; and Vol. 5: How Plants Defend Themselves.

587. _____. "Problems and Progress in Plant Pathology." **American Journal of Botany** 1 (March 1914): 97-111.

Links plant pathology to botany. Reviews the scientific opinions about the current status of plant pathology as a separate discipline.

588. Keitt, G.W. "History of Plant Pathology." In **Plant Pathology, An Advanced Treatise**, edited by J.G. Horsfall and A.E. Dimond, 61-97. New York: Academic Press, 1959.

Covers the period from 300 B.C. to the twentieth century. Begins with a discussion of botany and traces the transformation of plant pathology as a science.

589. Keitt, G.W., and F.V. Rand. "Lewis Ralph Jones, 1864-1945." **Phytopathology** 36 (January 1946): 1-17.

Brief review of the life of a founder of the American Phytopathological Society. Links his career to the development of plant pathology as a science in the United States. Includes an extensive bibliography of his publications.

590. McNew, G.L. "The Ever Expanding Concepts Behind Seventy-Five Years of Plant Pathology." Connecticut Agricultural Experiment Station, **Bulletin 663** (May 1963): 163-83.

History of plant pathology from the Dark Ages to the twentieth century. Emphasis is placed on the last seventy-five years of historical development.

591. Markham, Roy. "Landmarks in Plant Virology: Genesis of Concepts." **Annual Review of Phytopathology** 15 (1977): 17-39.

Discusses the major observations and ideas that have contributed to the development of plant virology as an interdisciplinary science. Brief review of each topic. Bibliography includes work in the U.S., Great Britain and Germany.

592. Morris, C.E., and P.C. Nicot. "Plant Pathology in the People's Republic of China: Getting the Job Done." **Plant Disease** 72 (August 1988): 648-60.

A brief history concerning research, institutions and teaching in China.

593. Porta-Puglia, A., M. Barba, and A. Quacquarelli. "L'Instituto Sperimentale per Patologia Vegetale: Presentazione, Organizzazione e Attivita." **Annali-dell'Instituto Sperimentale per la Patologia Vegetale** 11 (1986): 7-32.

A history of the Institute for Experimental Plant Pathology in Italy.

594. Pozaiak, Grazyna I. **Russian-English Dictionary of Helminthology and Plant Nematology** [**Russko-Angliiskii Slovar' po Gel'mintologii i Fitogel'mintologii**]. Farnham Royal, England: Commonwealth Agricultural Bureau, 1979. 108 pp.

See for terminology relating to worms and plant diseases.

595. Reed, George, M. "Phytopathology, 1867-1942." **Torreya** 43 (December 1943): 155-69.

A review of the classification of diseases caused by viruses and bacteria and the early methods used to control plant diseases.

596. Rodgers, Andrew Denny, III. "Erwin Frink Smith: A Story of North American Plant Pathology." **American Philosophical Society, Memoirs** 31 Philadelphia:

American Philosophical Society, 1952. 675 pp.

Smith became known as the dean of American plant pathologists because of his work in the USDA from 1886 to 1927. He helped professionalize plant pathology with his research on the bacterial causes of plant disease.

597. Schoen, J.F. "A Review of the Development of Seed Pathological Testing." **Journal of Seed Technology** 11 no. 2 (1987): 139-43.

A brief review of the major historical events in the development of seed pathology. Emphasizes testing techniques during the twentieth century. American and Danish scientific work is stressed.

598. Shear, Cornelius L. "First Decade of the American Phytopathology Society." **Phytopathology** 9 (1919): 165-70.

Very brief overview. Includes reprint of the report for the first meeting.

599. _____. "Plant Pathology, Retrospect and Prospect." **Science** 15 (April 1902): 601-612.

Discusses the advances made between 1880 and 1900. Includes a review of the concept that plants could change almost at will in relation to Darwin's theory of evolution.

600. Strobel, G.A. "Biological Control of Weeds." **Scientific American** 265 no. 1 (1991): 50-51, 54, 56D, 58-60.

A brief history of the use of pathogens for biological controls.

601. Thresh, J.M. "The Ecology of Tropical Plant Viruses." **Plant Pathology** 40 no. 3 (1990): 324-39.

Review of the role of British plant pathologists concerning tropical plant virology. Specifically notes work on virus diseases affecting cacao, rice, maize and banana crops.

602. Walker, J.C. **Plant Pathology**. New York: McGraw-Hill, 1969. 819 pp.

A general history of plant diseases. Useful for the history of agricultural science.

603. _____. "Some Highlights in Plant Pathology in the United States." **Annual Review of Phytopathology** 13 (1975): 15-29.

Discusses De Bary's contention in the nineteenth century that fungi were not distinct organisms or parasites on plants. Also includes a review of the work of Erwin F. Smith and L.R. Jones and the major plant blights in American agriculture.

604. Walker, J.C., and A.J. Riker. "Lewis Ralph Jones." **National Academy of Sciences, Biographical Memoirs** 31 (1958): 156-79.

Biographical article that emphasizes his life rather than analyzes his work. Includes a list of his publications. More scientific than Keitt's study.

605. Whetzel, H.H. **An Outline of the History of Phytopathology**. Philadelphia: W.B. Saunders, 1918. 130 pp.

A brief outline of the major developments in plant pathology from 500 B.C. to the early twentieth century.

606. Zadoks, J.C. "A Hundred and More Years of Plant Protection in the Netherlands." **Netherlands Journal of Plant Pathology** 97 no. 1 (1991): 3-4.

A history of Dutch plant pathology from 1890 to 1990. Discusses the introduction of chemotherapeutants during the 1940s and present-day environmental concerns.

607. Zoeten, G.A. de. "Risk Assessment: Do We Let History Repeat itself?" **Phytopathology** 81 no. 1 (1991): 585-86.

Notes the issues concerning genetically engineered protection from viruses. Warns of the dangers of releasing genetically engineered plants into the environment without adequate experimentation.

CORN

608. Alexander, D.E. "High-Oil Corn: Breeders Aim for Improved Quality." **Crops and Soils Magazine** 38 (August/September 1986): 11-12.

Brief note on the history of the feed value of corn. Makes the important point that high-oil corn produces more calories and that the science of nutrition has economic consequences.

609. Anderson, Edgar, and William L. Brown. "The History of the Common Maize Varieties of the United States Corn Belt." **Agricultural History** 26 (January 1952): 2-8.

Notes the origin and major types of dent and flint corn raised in the Midwest during the nineteenth century. Discusses early selection and breeding processes.

610. Becker, Stanley L. "Donald F. Jones and Hybrid Corn." Connecticut Agricultural Experiment Station, **Bulletin 763** (June 1976): 1-9.

Discusses the work of Jones in the development of hybrid corn and to convince geneticists, farmers and seed producers that hybrid corn merited their attention.

611. Bercaw, Louise O., Annie Hanney, and Nellie G. Larson. "Corn in the Development of the Civilization of the Americas: A Selected and Annotated Bibliography." United States Department of Agriculture,

Agricultural Economics Bibliography, No. 87, 1940. 195 pp.

A dated but useful bibliography. General topics include history, geography and botany.

612. Berlan, Jean-Pierre, and Richard Lewontin. "The Political Economy of Hybrid Corn." **Monthly Review** 38 (July/August 1986): 35-47.

Argues that claims about the social benefits of hybrid seed corn are "myths." Contends that yield increases due to hybrids are exaggerated but that hybrids created a perpetual market for seed corn.

613. Collins, G.N. "Notes on the Agricultural History of Maize." **Annual Report of the American Historical Association for the Year 1919**, 409-29. Vol. 1. Washington, D.C.: Government Printing Office, 1923.

Traces the origin of corn to human domestication in Mesoamerica. Should be read in conjunction with current publications on corn, especially those of Galinat and Mangelsdorf.

614. Coscia, A.A. **Desarrollo Maicero Argentino Cien Axos de Maiz en la Pampa**. Buenos Aires, Argentina: Hemisferio Sur., 1980. 120 pp.

Traces the production of maize in Argentina from 1875 to the present. Emphasizes economics but some discussion of technological influences since 1950.

615. Crabb, Alexander Richard. **The Hybrid-Corn Makers: Prophets of Plenty**. New Brunswick: Rutgers University Press, 1947. 331 pp.

An excellent study of the major contributors to the process of making hybrid corn. No one individual was responsible, but this study highlights the work of Donald F. Jones and his contribution to double-cross breeding.

616. Fedoroff, N.V., and Barbara McClintock. "The Restless Gene: How the Colors of Indian Corn Have Led to an Understanding of Wandering DNA." **Science New York** 31 no. 1 (1991): 22-28.

McClintock discusses her work since 1978 and her pioneering experiments on the transposable elements in maize.

617. Galinat, W.C. "The Origin of Corn." In **Corn and Corn Improvement**, edited by G.F. Sprague and J.W. Dudley, 1-31. Madison: American Society of Agronomy, 1988.

Discusses the origins of corn from teosinte to the modern hybrids. Stresses the importance of human action in the development of corn and discounts the the theory that corn developed in the wild.

618. Grilliches, Zvi. "Research Costs and Social Returns: Hybrid Corn and Related Innovation." **Journal of Political Economy** 66 (October 1958): 419-31.

Estimates the social rate of return for public and private funding for hybrid research. Concludes the return totalled at least 700 percent per year for each dollar invested in hybrid corn research by 1955.

619. Hallauer, A.R. "Corn Breeding." In **Corn Breeding**, edited by G.F. Sprague and J.W. Dudley, 463-564. **Agronomy Monograph No. 18**. Madison: American Society of Agronomy and Crop Science Society of America, 1988.

Traces the history of corn breeding. Detailed discussion of breeding methods, hybridization and improvements in genetic engineering.

620. _____. "The Methods Used in Developing Maize Inbreds." **Maydica** 35 no. 1 (1990): 1-16.

Reviews the history of breeding corn with emphasis on the selection of pedigree and inbred line development.

621. Hudson, M., E.J. LeRoux, and D.G. Harcourt. "Seventy Years of European Corn Borer (**Ostrinia nubilalis**) Research in North America." **Agricultural Zoology Reviews** 3 (1989): 53-96.

An extensive review of the research on this pest, including seasonal history, biological and chemical controls and breeding for host resistance.

622. Kiesselbach, T.A. "A Half Century of Corn Research: 1900-1950." **American Scientist** 39 (October 1951): 629-55.

A readable introduction to corn breeding. Considers both genetics and the scientists who made the major contributions.

623. Levine, E., and S.Y. Chan. **A Bibliography of the Northern and Western Corn Rootworms: An Update 1977 Through 1988**. Champaign: Illinois Natural History Survey, 1990. 29 pp.

Essentially limited to the chrysomelids **Diabrotica barberi**, **D. virgifera virgifera** and **D.V. zeae**. Excludes the literature that deals only with insecticide evaluations and recommendations.

624. Mangelsdorf, Paul C. **Corn: Its Origin, Evolution and Improvement**. Cambridge: Harvard University Press, 1974. 262 pp.

A general history of corn from its Mesoamerican origins to the late twentieth century. Good review of the major contributors to hybrid corn breeding and the technical processes to achieve hybridization.

625. Miles, Samuel R. "Performance of Corn Hybrids in Indiana, 1937-1946." Lafayette: Indiana Agricultural Experiment Station, **Bulletin 526**, 1948. 76 pp.

Hybrid corn had supplanted open-pollinated varieties by the late 1940s and produced 25 percent larger yields, but scientists had not yet developed varieties suitable for specific soils. Summarizes the performance of commercially available hybrids developed at university, experiment stations and USDA laboratories.

626. Mosher, Martin L. **Early Iowa Corn Yield Tests and Related Later Programs**. Ames: Iowa State University Press, 1962. 133 pp.

History of the corn demonstration and seed programs of Perry G. Holden. Summarizes the county demonstration work from 1904 to 1915, when corn production was based on the empirical selection of seed rather than genetics.

627. Paterniani, E. "Maize Breeding in the Tropics." **Critical Reviews in Plant Sciences** 9 no. 2 (1990): 124-54.

Discusses corn breeding methods and improvements, such as pest and disease resistance, nutritional quality and environmental adaptation.

628. Rhoades, M.M. "The Early Years of Maize Genetics." In **The Excitement and Fascination of Science: Reflections by Eminent Scientists**, edited by J. Lederberg, 819-47, Vol. 3. Palo Alto, CA: Annual Reviews, 1990.

The author, a specialist in the area of maize genetics, surveys the work of geneticists, such as Mendel, Correns, de Vries, Emerson, East, Harper, Bateson, Punnet, Stadler, Collins, Randolph and McClintock, during the early twentieth century.

629. Roepke, Howard G. "Changes in Corn Production of the Northern Margin of the Corn Belt." **Agricultural History** 33 (July 1959): 126-32.

Traces the shift from open pollinated to hybrid corn between 1940 and 1954. During that time, hybrid corn production increased from less than 33 percent to about 90 percent. In the corn belt, hybrid varieties comprised 97 percent of the corn grown. The pattern of hybrid adoption is shown by maps.

630. Russell, W.A. "Contribution of Breeding to Maize Improvement in the United States, 1920s-1980s." **Iowa State Journal of Research** 61 (August 1986): 5-34.

Since 1935 farmers have planted hybrid corn seed. Since that time breeders have sought to produce hybrids that will resist diseases, pests and lodging. Inbred lines also have been improved to enable commercial feasibility of single-cross hybrids. Contends that grain yields will continue to increase because of better breeding and cultural and management practices.

631. Sehgal, S. "The Contributions of World Maize Expert William L. Brown to Latin America." **Diversity** 7 no. 1-2 (1991): 43-44.

Summarizes Brown's work with corn in Latin America since 1940, including the collection and preservation of germplasm and breeding improvements. Notes his work to use tropical germplasm to improve corn in the Midwest of the United States.

632. Singleton, W. Ralph. "Early Researches in Maize Genetics." **Journal of Heredity** 26 (February 1935): 49-59 and (March 1935): 121-26.

Surveys the breeding experiments of scientists before the rediscovery of Mendel's law and the development of the agricultural experiment station system. Also discusses the technical processes used to obtain hybrid corn at the experiment stations during the late nineteenth century.

633. Timothy, David H., Paul H. Harvey, and Christopher R. Dowswell. **Development and Spread of Improved Maize Varieties and Hybrids in Developing Countries**. Washington, D.C.: Bureau for Science and Technology, Agency for International Development, 1988. 71 pp.

Maize ranks third behind rice and wheat as a cereal crop in the developing world. Research for the improvement of maize for those nations primarily is conducted at the International Maize and Wheat Improvement Center in Mexico and at the International Institute of Tropical Agriculture in Nigeria. Discusses the varieties and hybrids developed at these centers and evaluates their success.

634. Tollenaar, M. "Genetic Improvement in Grain Yield of Commercial Maize Hybrids Grown in Ontario from 1959-1988." **Crops Science** 29 no. 6 (1989): 1365-71.

During the late 1980s, Canadian agricultural scientists evaluated nine hybrids to explain the yield increase in machine-harvested maize. Genetic gains were credited for improving yields by reducing stem lodging and increasing plant density.

635. Wallace, Henry Agard, and William Lacy Brown. **Corn and Its Early Fathers**. East Lansing: Michigan State University Press, 1956. 141 pp.

Reviews the work on corn breeding from the eighteenth to the twentieth centuries. Includes discussion of the contributions of William James Beal, Robert and James Reid, George Krug, Isaac Hershey and Henry A. Wallace.

636. Watson, Stanley, and Paul E. Ramstad, ed. **Corn: Chemistry and Technology**. St. Paul: American Association of Cereal Chemists, 1987. 605 pp.

A technical study with world-wide scope. Discusses history, breeding, genetics and seed corn production and a host of commercial uses.

WHEAT

637. Ball, Carleton R. "The History of American Wheat Improvement." **Agricultural History** 4 (April 1930): 48-71.

Begins with the Spanish introduction of wheat to the West Indies in 1494. Discusses the use of lime to prevent rust in 1650 and colonial efforts to protect wheat against disease. Includes the introduction of new varieties during the nineteenth century and hybridization in the twentieth century.

638. Balla, L., Z. Bedo, and L. Lang. "The Aims and Results of Wheat Breeding in Hungary." **Sveriges Utsadesforenings Tidskrift** [Hungary] 99 no. 2 (1989): 109-16.

Discusses the major varieties in Hungarian grain production between 1951 and 1987. The average life span of a variety is five years. More than 50 percent of the varieties have been genetically improved, and breeders are working to develop frost and pest-resistant varieties as well as to improve the baking quality of flour.

639. Biswas, B.C., and R.K. Tewatia. "Wheat Production in India: Strategies for the Nineties." **Fertilizer News** 35 no. 12 (1990): 71-77.

Discusses the history of wheat production during the past twenty-five years. Contends that the Green Revolution in India was a wheat revolution. Attributes recent failures to increase yield to the inability of farmers to adopt the proper technology.

640. Bonjean, A., and E. Picard. **Les Cereales a Pialle: Origine, Histoire, Economie, Selection**. Paris: Softword, 1990. 205 pp.

A detailed description of the history of wheat, barley, rice, oats and rye and breeding practices. Designed for the introductory student.

641. Burdenyuk, L.A. "Methods and Results of Breeding Intensive-Type Varieties of Winter Wheat at the Belaya Tserkov' Breeding Station." **Selektsionnaya i Semenovodstvo Zernovykh i Zernobobovykh Kul'tur v Sisteme NPO "Sakhsvekla": Sbornik Nauchnykh Trudov** (1989): 12-19.

    A technical discussion of the varieties bred at this Ukranian experiment station since the 1950s.

642. Cai, Y.H. "Development and Production of Barley and Wheat in Shanghai, China P.R." **Rachis** 7 no. 1 (1988): 18-20.

    Describes the history of the varieties raised, farming procedures and environmental conditions near Shanghai.

643. Clark, J. Allen. "Improvement in Wheat." In **Yearbook of Agriculture, 1936**, 207-66. Washington, D.C.: United States Department of Agriculture, 1936.

    Discusses the improvements made by private and public agencies in the U.S. Maps show where specific varieties are grown. Some discussion of major scientists, diseases and international developments.

644. Dalrymple, D.G. "Changes in Wheat Varieties and Yields in the United States, 1919-1984." **Agricultural History** 62 (Fall 1988): 20-36.

Wheat yields did not increase substantially until the 1940s when biological science improved the development of hybrid varieties. Genetic changes enabled wheat varieties to combat diseases and to adapt to many environmental conditions. These scientific changes, together with the increased use of fertilizer, brought an agricultural revolution to wheat farmers in the United States.

645. Davis, Andrew. **Barberry Bushes and Wheat**. Cambridge: J. Wilson & Son, 1907. pp.

Reprinted from the publications of the colonial Society of Massachusetts, Vol. II. Discusses colonial agricultural practices and the discovery of the influence of the barberry on wheat rust. Good for historical background.

646. Dimitrov-Gotsov, K. "Collaboration of Bulgarian and Soviet Scientists in the Field of Plant Breeding." **Vestnik Sel'skokhozyaistvennoi Nauki Moskva** no. 8 (1989): 150-53.

Outlines cooperation in wheat breeding and the use of Russian varieties in Bulgaria.

647. Edwards, I.B. "New Horizons in Wheat Breeding." **North Dakota Farm Research** 45 no. 3 (November/December 1987): 30-34.

Discusses the major developments in spring wheat breeding in the U.S. between 1907 and 1962, including the effects of biotechnology.

648. He, M.S. "History and Future Prospects of Breeding Jin-type Wheat." **Scienta Agricultura Sinica** no. 2 (1985): 49-53.

Describes the wheat breeding activities at the Institute of Agricultural Science in Jinjang during the past twenty years. Twenty varieties are raised in ten provinces and autonomous regions. Spring wheats are the most popular.

649. Knudson, M.K. "The Invention and Diffusion of Two Competing Technologies: Semi-Dwarf and Hybrid Wheat." Ph.D. diss., University of Minnesota, 1988. 206 pp.

Publicly employed scientists developed and released the first semi-dwarf wheat in 1961, while privately employed scientists developed and released a new commercial hybrid wheat in 1983. Technological availability and adaptation enabled the rapid acceptance of semi-dwarf wheat by farmers while hybrid wheat seed was too expensive and the yields too undependable.

650. _____. "The Role of the Public Sector in Applied Breeding R&D: The Case of Wheat in the USA." **Food Policy** 15 (June 1990): 209-17.

Discusses the pros and cons of public-sponsored research, that is, government research. Uses semi-dwarf and hybrid wheats as an example of the appropriate public role. Contends that public-sponsored research is essential when private institutions fail to pursue work that does not have immediate economic value.

651. Lastovich, A.S. "Breeding Winter Wheat at the Ivanov Breeding and Experiment Station." **Selektsiya i Semenovodstvo Zernovykh i Zernobobovykh Kul'tur v Sisteme NPO "Sakhsvekla": Sbornik Nauchnykh Trudov** (1989): 96-99.

    Notes the history of wheat breeding at this Ukranian station since 1909. Hybridization began in 1951. Lists varieties.

652. Linhares, A.G., and J.L. Nedel. "Historico Sobre a Producao de Semente Genetica no Centro Nacional de Pesquisa de Trigo, Periodo 1975-1984." **Documentos Centro Nacional de Pesquisa de Trigo** no. 1 (1990): 24 pp.

    Discusses recent seed production at the National Wheat Breeding Centre in Brazil.

653. Malin, James C. **Winter Wheat in the Golden Belt of Kansas--A Study in Adaptation to Subhumid Geographical Environment**. Lawrence: University of Kansas Press, 1944. 290 pp.

    An excellent history of early wheat production in Kansas. This study will give historians of agricultural science an important perspective on wheat farming in the central Great Plains of the United States.

654. Martin, R.H., and W.J. Farrer. "In Farrer's Footsteps: Farrer Memorial Oration, 1980." **Journal of the Australian Institute of Agricultural Science** 47 no. 3 (1981): 123-31.

 Surveys wheat breeding for disease resistance in Australia during the twentieth century. Compares this research to the work of Farrer.

655. Morrison, J.W. "Marquis Wheat--A Triumph of Scientific Endeavor." **Agricultural History** 34 (October 1960): 182-88.

 Surveys the scientific work of Sir Charles Saunders, who developed Marquis wheat in 1892. This hard, red winter wheat resulted from the cross of Hard Red Calcutta and Red Fife in western Canada.

656. O'Brien, L. "Victorian Wheat Yield Trends, 1898-1977." **Journal of the Australian Institute of Agricultural Science** 48 no. 3 (1982): 163-68.

 Studies wheat breeding in Victoria, Australia, between 1898 and 1977 in relation to yields. Notes varieties.

657. Peters, Edmund A. "Joseph Danne: Oklahoma Plant Geneticist and His Triumph Wheat." **Chronicles of Oklahoma** 59 (Spring 1981): 54-72.

 Joseph Danne (1887-1959) developed the Triumph variety that dominates wheat production in the Southern Great Plains. A good biography of this self-taught, hybrid plant breeder.

658. Quisenberry, Karl S. "Let's Talk Turkey." **Heritage Review** 19 no. 4 (1989): 11-15. Brief discussion of the development of Turkey wheat. In 1872, the Mennonites introduced this variety of hard, red winter wheat to Kansas. Research began on this variety in 1874. Hybridization was accomplished in 1900.

659. _____, and L.P. Reitz. "Turkey Wheat: The Cornerstone of an Empire." **Agricultural History** 48 (January 1974): 98-110.

    A brief history of wheat growing in the winter wheat region of the Great Plains during the nineteenth century, especially among the Mennonites in Kansas. Discusses the major varieties that have been developed from Turkey wheat, and the relationship of these varieties to the Green Revolution.

660. Reitz, L.P. "The Improvement of Wheat." In **After a Hundred Years: Yearbook of Agriculture, 1962**, 108-15. Washington, D.C.: United States Department of Agriculture, 1962.

    Overview of wheat introductions and improvements in the U.S. Includes an account of breeding to develop disease-resistant varieties. A popular account, but useful to historians as an introduction.

661. Salmon, Samuel C., O.R. Mathews, and R. Leukel. "A Half Century of Wheat Improvement in the United States." **Advances in Agronomy** 5 (1953): 1-151.

A topical study that, in part, traces the scientific changes in wheat breeding to control diseases during the first half of the twentieth century. Includes the discussion of cultural practices and production.

662. Sapega, V.A. "Yield Variety Testing and Varietal Structure of Spring Wheat in Northern Kazakhstan." **Sibirskii Vestnik Sel'skokhozyaistvennoi Nauki** no. 2 (1990): 16-22.

Describes the varieties developed in Kazakhstan during the last thirty years. Contends that breeders have not produced high yield varieties capable of withstanding the harsh environment.

663. Schafer, John F. "Wheat Diseases and Disease Resistance in the Turkey Wheat Region." **Transactions of the Kansas Academy of Science** 77 (Fall 1974): 145-57.

Reviews the 100-year history of Turkey Red wheat on the Great Plains as well as wheat diseases and the development of resistant strains.

664. Schmidt, John W. "The Role of Turkey Wheat Germplasm in Wheat Improvement." **Transactions of the Kansas Academy of Science** 77 (Fall 1974): 159-72.

Discusses the wheat breeding program in Nebraska to develop disease resistant varieties and to improve yields.

665. Shepherd, James F. "The Development of Wheat Production in the Pacific Northwest." **Agricultural History** 49 (January 1975): 258-71.

Examines the natural, technical and economic factors that enabled wheat production in this region. Economic emphasis but a good introduction to the technology of wheat production.

666. _____. "The Development of New Wheat Varieties in the Pacific Northwest." **Agricultural History** 54 (January 1980): 52-63.

Traces the introduction of various wheat varieties and the breeding experiments of W.L. Spillman which began in 1899. An excellent article for the names of old and new varieties. Covers the period from the 1880s to the 1970s.

667. Sher, Mohammad. "Historical Background of Wheat Improvement in Baluchistan, Pakistan (1909-1980)." **Rachis** 8 no. 1 (1989): 10-12.

Surveys the improvement of wheat from 1909 to 1979. Notes varieties.

668. Simmonds, D.H. **Wheat and Wheat Quality in Australia**. Melbourne: CSIRO, 1989. 299 pp.

A history of wheat breeding and production in Australia.

669. Simpson, James. "La Eleccion de Tecnica en El Cultivo Triguero y El Atraso de la Agricultura Espanola a Finales del Siglo

XIX." **Revista De Historia Economica** 5 no. 2 (1982): 271-99.

Compares technological change in wheat production in Spain, Germany, France, Great Britain, Japan and the United States from 1880 to 1930. Emphasizes wheat production methods and technology in Spain during the late nineteenth century.

670. Socolofsky, Homer E. "The World Food Crisis and Progress in Wheat Breeding." **Agricultural History** 43 (October 1969): 423-37.

Traces the improvements in corn, wheat and grain sorghum to enable production in almost any country and under a variety of environmental conditions. Notes the lag in genetic research in the Soviet Union and its effects on grain production.

671. Stoppler, H., E. Kolsch, and H. Vogtmann. "Auswirkungen der Zuchtung bei Wintererweizen in einem Landwirtschaftlichen System mit Geringer Betriebsmittelzufuhr von Aussen." **Journal of Agronomy and Crop Science** 162 no. 5 (1989): 325-32.

Discusses the twenty-four varieties of wheat developed in the Federal Republic of Germany between 1900 and 1983. Notes productivity.

672. Waddington, S.R., et al. "Improvement in the Yield Potential of Bread Wheat Adapted to Northwest Mexico." **Crop Science** 26 no. 4 (1986): 698-703.

A study of wheat yields per hectare for cultivars released in Mexico between 1950 and 1982. Concludes that wheat varieties planted after 1970 had 16 percent greater biomass and increased productivity.

673. _____. "The Yield of Durum Wheats Released in Mexico Between 1960-1984." **Journal of Agricultural Science** [Great Britain] 108 no. 2 (1987): 469-77.

Discusses the scientific experiments with eight cultivars at the Instituto Nacional de Investigaciones Agricolas. This work has improved the yield of durum wheats in northwest Mexico during the last twenty-five years.

674. Wang, N.L. "Changes of Wheat Cultivars in Zhejiang During 1935-86." **Zhejiang Agricultural Science** no. 5 (1987): 201-07.

A brief history of wheat breeding in the Zhejiang province of China.

675. Weber, A. "Stand und Entwicklung der DDR-Agrarproduction." **FS Analysen** no. 5 1 (1989): 26-38.

Compares agriculture in the German Democratic Republic and the Federal Republic of Germany with emphasis on wheat, sugar beets and potatoes during the last century.

676. Wessel, Thomas R. "Wheat for the Masses: M.L. Wilson and the Montana Connection." **Montana, the Magazine of Western History** 31 no. 2 (1981): 42-53.

Discusses the work of the Russians to adapt the dry land wheat farming techniques developed in Montana by Wilson. He visited the Soviet Union during the 1920s to assist with the adoption of his techniques. The government prohibited decision making by farm managers and this system of wheat farming failed.

677. Zeven, A.C. "Landraces and Improved Cultivars of Bread Wheat and Other Types Grown in the Netherlands Up to 1944." **Agricultural University Wageningen Papers** no. 2 (1990): 103 pp.

Lists old Dutch names for wheat varieties and provides descriptions of varieties raised during the seventeenth and eighteenth centuries; notes the historical literature of wheat diseases and identifies collections of historical varieties.

RICE

678. Alvarez, J., G.H. Snyder, and D.B. Jones. "The Integrated Program Approach in the Development of the Florida Rice Industry." **Journal of Agronomic Education** 18 no. 1 (Spring 1988): 6-10.

The emergence of rice farming in Florida resulted from economic need for diversification. Technology in the form of combines and dryers and science in the form of seed improvement enabled profitable production. By the late 1980s, rice had become a viable farming alternative.

679. Anderson, Robert S., Edwin Levy, and Barrie M. Morrison. **Rice Science and Development Politics: Research Strategies and IRRI's Technologies Confront Asian Diversity, 1950-1980.** Oxford: Clarendon Press, 1991. 394 pp.

A history of research on rice and the International Rice Research Institute. Discusses technological transfer in Sri Lanka and Bangladesh.

680. Bray, Francesca. **The Rice Economies: Technology and Development in Asian Societies**. New York: B. Blackwell, 1986. 254 pp.

A history of rice technology in relation to social and economic change. Concentrates on East and Southeast Asia. Good background on water control, labor and capital and the origins of rice. Uses China and Japan as historical examples.

681. Campell, Joseph K. **Dibble Sticks, Donkeys, and Diesels: Machines in Crop Production**. Manila: International Rice Research Institute, 1990. 329 pp.

A history of agricultural technology in rice production, with an emphasis on Asia.

682. Camona, P.S. "Melhoramento de Arrozo Irrigado na Regiao sul do Brasil." **Lavoura Arrozeira** [Brazil] 42 (1989): 16-16.

Surveys rice breeding in the states of Rio Grande do Sul and Santa Catarina in

Brazil. Emphasis on varieties resistant to rice blast and iron toxicity.

683. Chandler, R.F., Jr. **An Adventure in Applied Science: A History of the International Rice Research Institute**. Los Banos, Laguna, Philippines: International Rice Research Institute, 1982. 233 pp.

Traces the origin, development and research program of the Institute during the past twenty years. This center pioneered in the area of rice research and helped revolutionize Asian agriculture.

684. Chen, Y.W. "Hybrid Rice in China." **Journal of Hunan Science and Technology** [China] 1 nos. 2-3 (1985): 1-9.

A review of hybrid rice breeding in Hunan since 1972. The first rice hybrids were made available to farmers in 1976. Yields have improved by 30 percent over conventional rice varieties. In some areas where the growing conditions are unfavorable to conventional varieties, yields have improved by 100 percent with hybrid seed.

685. Cho, K. "Factors Affecting the Changes in Utilization of Paddy-Field in Tokugawa Era." **Science Bulletin of the Faculty of Agriculture, Kuyshu University** [Japan] 40 nos. 2-3 (1986): 75-109.

Describes paddy-field rice farming from 1600 to 1868 and discusses the double-cropping system with the use of irrigation technology.

686. Dethloff, Henry C. **A History of the American Rice Industry, 1685-1985**. College Station: Texas A & M University Press, 1988. 215 pp.

A brief history with an economic focus, but it provides some discussion of technological change. A good introduction to the rice industry in the United States.

687. Guimaraes, E.P., and R. de P. Ferreira. "Origen, Evolucion, y Resultados de la Red Basilena de Evaluacion de Germoplasma de Arroz." **Arroz en las Americas** 9 no. 2 (1988): 1-3.

Outlines the history of rice breeding in Brazil. Emphasizes the work of the National Centre for Rice and Bean Research as a coordinator of research groups in the various Brazilian states.

688. Hargrove, T.R., V.L. Cabanilla, and W.R. Coffman. "Twenty Years of Rice Breeding." **BioScience** 38 no. 10 (November 1988): 675-81.

Discusses the role of semi-dwarf varieties in rice breeding in Asia. Contends that because most Asian rices contain similar cytoplasm they are vulnerable to disease and blight that could destroy the crop across the region.

689. Khush, G.S. "Rice Breeding: Accomplishments and Challenges." **Plant Breeding Abstracts** 60 no 5. (1990): 461-69.

Reviews the major achievements in rice breeding during the last twenty-five years. Since 1965, rice production has increased 82 percent. Improved varieties are raised on 65 percent of the world's rice lands.

690. Stewart, M.A. "Rice, Water, and Power: Landscapes of Domination and Resistance in the Low Country, 1790-1880." **Environmental History Review** 15 (Fall 1991): 47-64.

Discusses rice cultivation and irrigation in the coastal areas of Georgia and South Carolina in reference to the ecology.

691. Suge, Hiroshi. **Shonai ni Okeru Suito Minkan Ikushu no Kenkyu.** Tokyo: No-san-gyoson Bunka Kyokai, 1990. 318 pp.

Discusses local traditional rice breeding in the Shonai Region. See for a general history of rice farming in this region.

692. U, Khin Win. **A Century of Rice Improvement in Burma**. Manila: International Rice Research Institute, 1991. 162 pp.

A history of hybrid rice breeding in Burma. See the bibliography for additional sources.

693. Van, T.K. "History of Plant Breeding for Seed Improvement in Padi." In **Seed Technology in the Tropics**, edited by H.F. Chin, I.C. Enoch, and R.M.R. Harum, 13-20. Serdang, Malaysia: Universiti Pertanian Malaysia, 1977.

A brief review of the origin and development of rice breeding in Malaysia.

694. Win, K. "Myanmar's Experience in Rice Improvement, 1830-1985." **IRRI Research Paper Series, No. 141**. Manila: The Institute, 1990. 10 pp.

Brief description of rice breeding in Myanmar (Burma).

695. Yu, L.Q. "A Comment on the Origin of Cultivated Rice After Reading 'A History of Agriculture in India.'" **Acta Agronomica Sinica** 15 no. 1 (1989): 86-93.

Reviews the bibliography of rice production, with emphasis on the works of Randhawa, Vavilov and de Candolle. Contends that contrary to Vavilov rice cultivation began in China not India.

696. Yukawa, K. "Historical Development of Rice Cultivation and Its Water Management." **Journal of Irrigated Engineering and Rural Planning** no. 16 (1989): 60-70.

Surveys the history of rice cultivation and water management during the last 2,000 years.

SOYBEANS

697. Belson, Abby Avin. "Super Soy." **Family Health** 11 (October 1979): 52-53, 55.

Outlines the use of soybeans for human consumption in China, Japan and the United States. Notes the use of soybeans in medical research.

698. Craigmiles, J.P, et al. "Advances in Soybeans: History and Technology." Texas Agricultural Experiment Station, **Bulletin 11** (April 1981): 5-9.

Brief discussion of the origins of soybeans in East Asia and research developments to increase yields and decrease disease. The remainder of the bulletin emphasizes soybean development along the coastal prairie of Texas.

699. Hymowitz, T. "Soybeans: The Success Story." In **Advances in New Crops. Proceedings of the First National Symposium "New Crops Research, Development, Economics," Indianapolis, IN, October 1988,** edited by J. Janick and J.E. Simon, 23-26. Portland, OR: Timber Press, 1990.

Surveys the history of soybeans from the origin of the crop in China. Notes its adoption in the United States in 1765. Primarily a forage crop in the U.S. until the 1920s. Discusses germplasm collection in China, Japan and Korea during the 1920s and 1930s.

700. Miguel Gordillo, E. de, F. Perez-Gragera, and D.I. Libran-Garcia. "La Soja: Una Alternativa para los Regadios Extremenos." **Agricultura Madrid** 59 (1990): 50-53.

Outlines the history of soybeans in Asia, Europe and the Untied States.

701. Piper, C.V. "The Name of the Soybean: A Chapter in its Botanical History." **Journal of the American Society of Agronomy** 6 (March/April 1914): 75-84.

Discusses the varieties and taxonomy of soybeans since 1794.

702. Smith, A.K., and Sidney J. Circle, eds. **Soybeans: Chemistry and Technology.** Vol. 1, **Proteins**, Westport, CT: AVI Publishing Company, 1978. 470 pp.

See chapter 1 by the editors for a discussion of the history of soybean production and uses in Asia and the United States.

703. Specht, J.E., and J. H. Williams. "Contribution of Genetic Technology to Soybean Productivity--Retrospect and Prospect." In **Genetic Contributions to Yield Grains of Five Major Crop Plants: Proceedings of a Symposium Sponsored by Div. C-1, Crop Science Society of America 2 December 1981, in Atlanta Georgia**, edited by W.R. Fehr, 49-74. Madison: Crop Science Society of America, 1984.

A technical study with some historical value on the genetics and productivity of soybeans in the United States.

704. Windish, Leo G. **The Soybean Pioneers: Trailblazers, Crusaders, Missionaries**. Galva, IL: The Author, 1981. 239 pp.

Traces the growth and development of soybean production in the United States. Includes technological change and background on the major associations, companies, laboratories and scientists that contributed to soybean production.

FEED AND FORAGE

705. Ball, Carleton R. "The History and Distribution of Sorghum." United States Department of Agriculture, Bureau of Plant Industry, **Bulletin 175**, 1910. 63 pp.

Discusses the origin and history of sorghum. Describes the varieties of sorghum in various geographic regions of Africa, Europe, South America and North America. Notes independent development in Africa and India. Economic rather than scientific or technological in focus.

706. Beverdam, H.J. "A Focus on the Quality Aspects of Grass Seed Production." **Prophyta** 6 (1989): 8, 19, 21-23.

An overview of the breeding of forage grasses in the Netherlands. Notes the work of the Dutch Certifying Agency regarding seed testing.

707. Chapline, William R., et al. "The History of Western Range Research." **Agricultural History** 18 (July 1944): 127-143.

A comprehensive program of range research was not established until the mid-1930s, when soil erosion problems compelled action to save the western grasslands. Emphasizes the vegetation experiments and related work of the USDA. Concludes with the influence of World War II.

708. Coffman, F.A. "Origin of Cultivated Oats." **Journal of the American Society of Agronomy** 38 (November 1946): 983-1002.

Traces modern cultivated oats to wild red oats. Although the narrative is heavy with scientific language, this study is useful for tracing the genetic and physiological records.

709. Erf, Oscar. "History of Modern Hay Making." **Jersey Bulletin** 57 pt. 1 (April 6, 1938): 451-52, 478-79; pt. 2 (April 13, 1938): 495, 520-21.

Discusses the vitamin and mineral content of hay as well as the harvesting and curing processes.

710. Green, B.H. "Agricultural Intensification and the Loss of Habitat, Species and Amenity in British Grasslands: A Review of Historical Change and Assessment of Future Prospects." **Grass and Forage Science** 45 no. 4 (1990): 365-72.

Without management British grasslands would revert to forest. Discusses the need to balance agricultural and environmental concerns with scientific management.

711. Huckabay, John Porter. "The Origin of Cultivated Sorghum." Oklahoma State University, Ph.D. diss., 1967. 91 pp.

Argues that sorghum was domesticated in Northeastern Africa in the region of present-day Ethiopia about 5,000 years ago. This crop then spread into Asia Minor, India and Southeast Asia and to southern and western Africa. Farmers continually modified sorghum through the process of selection and breeding to adapt it to local conditions.

712. Kramer, N.W. "Grain Sorghum Production and Breeding--Historical Perspectives to Future Prospects." **Report of the Annual Corn and Sorghum Research Conference** 42 (1987): 1-9.

Essentially economic in emphasis but some discussion of technology, hybrids and germplasm. Stresses the period from 1930 to the early 1980s.

713. Lamson-Scribner, F. "Progress of Economic and Scientific Agrostology." In **Yearbook of Agriculture, 1899**, 347-66. Washington, D.C.: United States Department of Agriculture, 1899.

Traces the scientific study of grasses and forage plants for agricultural purposes from 1878 to the creation of the USDA's Division of Agrostology in

1895. Brief descriptions of the most important grasses and forage plants. Notes the most significant work of the USDA.

714. Marlon, P.J. "Improving Productivity in Sorghum and Pearl Millet in Semi-arid Africa." **Food Research Institute Studies** 11 no. 1 (1990): 1-43.

Reviews the inability of researchers to improve yields in this portion of West Africa. Contends that more emphasis needs to be placed on new technologies that will be adaptable to the environmental limitations for agriculture in the region.

715. Piper, Charles, V., and Katherine S. Bort. "The Early Agricultural History of Timothy." **Journal of the American Society of Agronomy** 7 (January-February 1915): 1-14.

Traces the discovery of timothy to the northern United States, perhaps in New Hampshire. Named for Timothy Hanson who introduced this grass to England and helped disseminate it in the colonies during the mid-eighteenth century.

716. Rekunen, J. "Advances in the Breeding of Oats: Comparative Treats with Historical Varieties in 1977-87." **Journal of Agricultural Science in Finland** 60 no.4 (1988): 307-21.

Evaluates oat breeding in Finland. Concludes that the improvements in breeding between 1921 and 1982 increased yields approximately 40 percent and decreased lodging problems in harvesting equipment.

717. Sampson, D.R. "On the Origin of Oats." Harvard University **Botanical Museum Leaflet**, 16 (November 26, 1954): 265-303.

Presents a description and an identification of the habitat of various varieties. Traces the oldest known oat grains to Egypt in the twelfth century. Notes the efforts of oat breeders to develop fungi-resistant plants.

718. Schukking, S. "The History of Silage Making." **Stikstof** no. 19 (1976): 2-11.

A brief history from ancient times to the 1970. Discusses the technical requirements and chemical changes in the silage making process. Some discussion of technological change. Focus on the Netherlands.

719. Tabor, Paul. "The Early History of Annual Lespediza in the United States." **Agricultural History** 35 (April 1961): 85-89.

Introduced from China or Japan during the mid-nineteenth century. After the Civil War, this clover became a major forage crop, particularly in the South.

720. Wasser, C.H. "Early Development of Technical Range Management, 1895-1945." **Agricultural History** 51 (January 1977): 63-77.

During this time, range management became a science for the control of forage grasses and grazing. The creation of a new Division of Agrostology in the USDA played a major role in the professionalization of this work and related research.

721. Williams, Thomas A. "Succulent Forage for the Farm and Dairy." In **Yearbook of Agriculture, 1899**, 613-26. Washington, D.C.: United States Department of Agriculture, 1899.

An economic rather than a scientific discussion of the primary crops useful for feeding dairy cows, such as corn, clover, millets, legumes, sorghums, oats, corn and cabbage.

722. Youngberg, H.W., et al. "Grass and Legume Seed Production." In **Forages: The Science of Grassland Agriculture**, edited by M.E. Heath, R.F. Barnes and D.S. Metcalf, 72-79. Ames: Iowa State University Press, 1985.

Reviews the history with emphasis on production, diseases, pests, storage and marketing in the United States.

COTTON

723. Afzal, I. "The Classification of Pakistan Cotton." **Textile Asia** 20 no. 8 (1989): 171-74.

Traces the history of cotton cultivation in Pakistan, notes the current status of cotton production in relation to the world market.

724. Faure, G., K. Djagni, and R. Raymond. "Trois and de Recherche Agro-Economiques dans la Zone Cotonniere Togolaise." **Coton et Fibres Tropicales** 43 no. 2 (1988): 101-10.

Reports the recent research on cotton production in Togo. Economic and social rather than scientific in emphasis.

725. Fulmer, John L. **Agricultural Progress in the Cotton Belt Since 1920**. Chapel Hill: University of North Carolina Press, 1950. 236 pp.

Primarily economic in emphasis but helpful for studying the use of tractors and the efforts to mechanize the cotton harvest.

726. Hau, B. "History of Cotton Plant Breeding in Cote-d'Ivoire." **Coton et Fibres Tropicales** 43 no. 3 (1988): 177-204.

Traces the beginning of cotton production in Africa to the seventeenth century in the Ivory Coast, where it thrived because of its resistance to pests. Since 1960, herbicides and hybrid varieties have made cotton the preeminent crop of the savanna region. Notes varieties raised.

727. Kohel, R.J., and C.F. Lewis, ed. **Cotton**. Madison: American Society of Agronomy, 1984. 605 pp.

Contains fourteen essays on topics such as the origin of cotton, germplasm resources, morphology, genetics, breeding, harvesting and ginning practices. Useful historical summaries.

728. Lancon, J., J.L. Chanselme, and C. Klassou. "Bilan du Progres Genetique Realise par la Recherche Cotonniere au Nord-Cameroun de 1960 a 1988." **Coton et Recherche du Coton et des Textiles Exotiques, IRCT-CIRAD** 45 no. 2 (1990): 145-67.

A technical review of the eight varieties developed between 1951 and 1987. Some historical value for breeding developments.

729. Moore, John Hebron. "Cotton Breeding in the Old South." **Agricultural History** 30 (July 1956): 95-104.

Surveys the cotton industry prior to the adoption of Mexican hybrid seed, a variety superior to Creole black seed and Georgia green seed cotton. Includes a discussion of cotton breeding in the Lower South and the work of Rush Nutt and Henry W. Vickle.

730. Paddock, F.B. **The Cotton or Melon Louse: Life History Studies**. College Station, Texas Agricultural Experiment Station, **Bulletin 257**, 1919. 54 pp.

The cotton or melon louse was first discovered in Texas during the late nineteenth century. Because the louse attacked melons, few scientists and farmers thought that it damaged cotton plants. Scientists did not begin to study the cotton louse until 1916. This publication essentially discusses its life history and economic importance.

731. Percival, A.E., and R.J. Kohel. "Distribution, Collection, and Evaluation of Gossypium." **Advances in Agronomy** 44 (1990): 225-56.

Brief description of the history of domestication, distribution and collection of germplasm by the U.S. National Cotton Germplasm Collection at College Station Texas. Includes a discussion of plant explorations since 1983 and genetics.

732. Ruf. T. "Deux Siecles d'Interventions Hydrauliques et Cotonniere dans la Valle du Nil." **Dynamique des Systemes Agraires** 2 (1985): 279-310.

A survey of irrigation and cotton production in the Nile Valley during the past two centuries.

733. Stephens, S.G. "The Origins of Sea Island Cotton." **Agricultural History** 50 (April 1976): 391-99.

Sea Island cotton resulted from the hybridization of domesticated and wild cotton from South America. Planters raised this variety from the late nineteenth to the early twentieth century.

734. Street, James H. **The New Revolution in the Cotton Economy**. Chapel Hill: University of North Carolina Press, 1957. 294 pp.

An essential introduction to the mechanization of cotton farming. Economic in emphasis but good for technological developments.

CHAPTER XI

ANIMAL SCIENCES

GENERAL

735. **Actes: 3eme Congres Mondial de Reproduction et Selection des Ovins et Bovins a Viande**. 2 Vols. Paris: Institut National de la Recherche Agronomique, 1988.

     Proceedings of the 3rd World Congress on Sheep and Cattle Breeding. Good for a contemporary account of current breeding practices. Discusses modern techniques of genetic improvement, technological innovations and problems to be solved.

736. Ammerman, C.B., and P.R. Henry. "Ruminant Nutrition: A Century of Progress." **Cornell Veterinarian** 75 (January 1985): 174-90.

     Scientists began to make progress in mineral nutrition for ruminant animals during the 1930s. By the mid-1980s, they had identified 21 mineral elements essential for good animal nutrition.

They also discovered the importance of Vitamins A, D and E and improved low value forages with chemical treatments to increase weight gains.

737. Barton, R.A. "The Role of Breed Societies: Past, Present and Probable." **Proceedings of the 2nd World Congress on Sheep and Beef Cattle Breeding, 16-19 April 1984**, 177-85. Pretoria, South Africa, South African Stud Book and Livestock Improvement Association, 1984.

A look at the past, present and future role of breed societies. In today's environment breed societies will need to work with agricultural scientists to adopt new technology. Breed societies may need to consolidate in order to afford new technology.

739. Becker, Stanley Leonard. "The Emergence of a Trace Nutrient Concept Through Animal Feeding Experiments." Ph.D. diss., University of Wisconsin, 1968. 377 pp.

Discusses the history of the trace nutrient concept from 1840 and the development of chemical analysis. The work at the Connecticut and Wisconsin experiment stations is emphasized in relation to the discovery of chemicals that are necessary daily in small quantities to ensure growth and good health. Important for the historical study of animal nutrition.

740. Byerly, T.C. "Changes in Animal Science." **Agricultural History** 50 (January 1976): 258-274.

Discusses the work of the Bureau of Animal Industry in the USDA and its successor agencies. Emphasizes research in the areas of nutrition, breeding and chemical and ecological sciences relating to the livestock and poultry industry.

741. Cook, P.W. "Common and Preferred Stock: Neat Cattle, Coarse Sheep and Fat Hogs, 1620-1800." In **Exploring Our Livestock Heritage: Proceedings of the 1988 Annual Meeting of the American Minor Breeds Conservancy**, 13-25. Pittsboro, NC: The Conservancy, 1988.

A general discussion of breeding and raising cattle, sheep and swine in New England.

742. Evans, J. Warren, and Alexander Hollaender, ed. **Genetic Engineering of Animals: An Agricultural Perspective**. New York: Plenum Press, 1986. 328 pp.

Discusses the current status of animal bioengineering and the application of genetic technology to food production. Includes the discussion of the moral and legal implications of genetic engineering of animals as well as essays on genetic manipulation, bovine growth hormones, gene mapping and the history of genetic engineering of farm animals.

743. Klickstein, Herbert S. "Charles Caldwell and the Controversy in America Over Liebig's 'Animal Chemistry.'" **Chymia** 4 (1953): 129-57.

In 1845, Caldwell attacked Liebig's theories and scientific approach in general. This article reprints many of Caldwell's lectures, writings and speeches while at the Louisville Medical Institute and the University of Pennsylvania.

744. Legates, J.E. "Developments in Theory and Application of Animal Breeding." **Proceedings of the World Symposium in Honor of Professor R.D. Politiek, Organized by the Agricultural University, Wageningen, Netherlands, 11-14 September 1988**, 15-26. Wageningen, Netherlands: Center for Agricultural Publishing and Documentation, 1988.

A survey of animal breeding and selection during the last 200 years.

745. Mason, I.L. **A World Dictionary of Livestock Breeds, Types and Varieties**. 2nd ed. Farnham Royal, England: Commonwealth Agricultural Bureau, 1969. 268 pp.

First published in 1951, this dictionary provides a basic description and background for breeds of domestic livestock.

746. Oltjen, Robert R. "Significant Milestones in the 75 Year History of the American Society of Animal Science." **Journal of Animal Science** 57 Supplement 2 (July 1983): 1-15.

Notes the four major historical events of the Society: (1) founding in 1908; (2) changing emphasis from nutrition and production to broad-based science; (3) start of the **Journal** in 1942; and (4) focus on the scientific aspects of animal science in 1963.

747. Phillips, Ralph W. "Breeding Better Livestock." In **Yearbook of Agriculture: Science in Farming, 1943-1947**, 33-60. Washington, D.C.: United States Department of Agriculture, 1947.

Discusses the processes of selection, inbreeding and cross-breeding as tools for genetic improvement, particularly in relation to swine and sheep. Brief survey of the development of new breeds.

748. Rexroad, C.E. Jr. "History of Genetic Engineering of Laboratory and Farm Animals." In **Genetic Engineering of Animals: An Agricultural Perspective**, edited by J. Warren Evans and Alexander Hollaender, 127-138. New York: Plenum Press, 1986.

An excellent place to begin research on this subject.

749. Russell, N. **Like Engend'ring Like: Heredity and Animal Breeding in Early Modern England**. Cambridge: Cambridge University Press, 1986. 271 pp.

A history of breeding cattle, sheep and horses in England from the sixteenth to the eighteenth century. Useful for those who are not scientists, because technical terms for breeding and genetics have not been used. Emphasizes description of breeds and breeding practices.

750. Warwick, Everett J. "New Breeds and Types." In **After A Hundred Years: Yearbook of Agriculture 1962**, 276-81. Washington, D.C.: United States Department of Agriculture, 1962.

Brief introduction to traditional American breeds of swine, sheep and cattle. Includes new breeds as well as those that essentially disappeared after 1850.

751. Willham, R.L. "From Husbandry to Science: A Highly Significant Facet of Our Livestock Heritage." **Journal of Animal Science** 62 (June 1986): 1742-58.

Examines the contributions of husbandmen and scientists to livestock development. Emphasizes economic and institutional factors rather than science.

BEEF CATTLE

752. Allen, Lewis F. **American Cattle: Their History, Breeding and Management**. New York: Taintor Brothers, 1868. 528 pp.

A general, topical history of cattle, including breed improvement, diseases and care. Dated but still of value.

753. Baker, C.M.A., and C. Manwell. "The Breton Breed of Cattle in Britain: Extinction Versus Fitness." **Agricultural History Review** [Great Britain] 35 pt. 2 (1987): 171-78.

Discusses the history of this dairy cattle breed in the United Kingdom.

754. Benyshek, L.L., and J.K. Bertrand. "National Genetic Improvement Programs in the United States Beef Industry." **South African Journal of Animal Science** 20 no.3 (1990): 103-09.

Describes the historical development of procedures to evaluate cattle breeding since 1950. Mathematical in methodology.

755. Boezio, S. "Criollo Cattle and Sheep in Uruguay." In **Genetic Conservation of Domestic Livestock**, edited by L. Anderson, 108-11. Wallingford, United Kingdom: CAB International, 1990.

Brief history of Criollo cattle and sheep with notes about current production.

756. Johnson, Charles Wilford. "Background of Major Changes in the Beef Industry." **Journal of Geography** 64 (February 1965): 53-59.

Discusses the physical changes of cattle that are necessary to permit their adaptation to grassland feeding rather than corn feeding. Contends that present-day cattle are not capable of efficiently utilizing large quantities of roughage. Urges the increased use of Brahman cattle.

757. Leavitt, Charles T. "Attempts to Improve Cattle Breeds in the United States, 1790-1860." **Agricultural History** 7 (April 1933): 51-67.

Emphasizes the introduction of Shorthorn cattle. Surveys the importance of the agricultural fairs and the press in promoting breeding improvements.

758. McKillop, R.F. "Village Beef Cattle Development: The Melanesian Experience." **Islands/Australia Working Papers, No. 89/2**, National Centre for Development Studies, Australian National University, 1989. 31 pp.

Discusses the programs to introduce cattle into the village agriculture of Papua New Guinea, Solomon Islands, Vanuatu, and Fiji. Contends that government agencies have provided ineffective support.

759. Malin, Donald F. **The Evolution of Breeds: An Analytical Study of Breed Building in Shorthorn, Hereford, and Aberdeen Angus Cattle, Poland China and Duroc Jersey Swine**. Des Moines, Wallace Pub. Co., 1923. 278 pp.

Describes the history and pedigree of these major breeds. Dated but useful.

760. Maree, C. "South African Beef Industry: Cows, Calves and Grass." **Beef Cattle Science Handbook** 22 (1987): 259-61.

Outlines the history of breeds in South Africa.

761. Maule, J.P. **The Cattle of the Tropics**. Edinburgh, Scotland: University of Edinburgh Centre for Tropical Veterinary Medicine, 1990. 225 pp.

Surveys the history of indigenous tropical cattle. Discusses breeds and breeding in Asia, Africa, South America, Australia and the Caribbean. Includes old-established and synthetic breeds.

762. Middleton, B.K., and J.B. Gibb. "An Overview of Beef Cattle Improvement Programs in the United States." **Journal of Animal Science** 69 (September 1991): 3861-3871.

Historical review with emphasis on technological developments as well as organizations and financing. Notes that the associations provide the primary information on performance programs. Some discussion of the technical problems confronting the industry in the United States.

763. Rhoad, Albert O., ed. **Breeding Beef Cattle for Unfavorable Environments**. Austin: University of Texas Press, 1955. 248 pp.

Discusses the effects of the environment on animal health and reproduction, the relationship of soil fertility and nutrition to cattle production and the improvement of range grasses in hot regions. Also emphasizes the necessity of adapting cattle to their environment. Includes breeding reports from scientists in South Africa, Brazil and the United States and the use of native Afrikander, Hereford, Gyr, Zebu, Santa Gertrudis and other breeds.

764. Rifkin, Jeremy. **Beyond Beef: The Rise and Fall of the Cattle Culture**. New York: Dutton, 1992. 353 pp.

A general history of the cattle industry in the United States. This study primarily will be of value to historians of agricultural science for its discussion of the environmental aspects of cattle raising.

765. Rodrigues da Cunha, J.G. "Zebu Cattle Around the World." **Beef Cattle Science Handbook** 22 (1988): 252-58.

A brief history of Zebu cattle including breed strains and programs. Emphasizes Brazil.

766. Sanders, Alvin H. **A History of Aberdeen-Angus Cattle**. Chicago: New Breeder's Gazette, 1928. 1,042 pp.

Essentially a reference tool for breeders concerned with lineage and historians interested in the origin of the breed in England. Includes descriptions of the major breeders and importers.

767. _____. **Short-Horn Cattle: A Series of Historical Sketches**. Chicago: Sanders Publishing Co., 1918. 1,021 pp.

Traces the history of Shorthorns in the United States from the British colonial period to the early twentieth century. Emphasizes breed improvements after the mid-nineteenth century.

768. Smith, G.F. "Four Decades of Technical Progress in Cattle Artificial Insemination." **Veterinary History: Bulletin of the Veterinary** [Great Britain] 5 (Winter 1987/1988): 59-70.

A history of artificial insemination since the 1930s. The author attributes the earliest work in this area to Russian breeders. From Russia, the techniques for artificial insemination spread to other countries, particularly Denmark, the United States and the United Kingdom.

769. Sweet, Orville K. **Birth of a Breed: The History of Polled Herefords, America's First Beef Breed**. Kansas City, MO: Lowell Press, 1975. 386 pp.

A general history that includes descriptions of important individuals associated with Polled Hereford breeding in the United States. Economic and institutional in nature, but some information on breed improvement and genetics.

770. Takayasu, I. "Studies on the Establishment and History of Improvement of the Japanese Shorthorn Breed of Cattle." **Bulletin of the Faculty of Agriculture**, [Hirosaki University] no. 40 (1983): 37-108.

A detailed discussion and a good introduction to cattle breeding in Japan.

771. Thompson, James Westfall. **A History of Livestock Raising in the United States, 1607-1860**. Agricultural History Series No. 5. Washington D.C.: United States Department of Agriculture, 1942. 182 pp.

A general economic history, but useful for background on early breeds and introductions to American agriculture.

772. Willham, R.L. "Information-Based Breeding." In **Emerging Technology and Management for Ruminants: 1985 Livestock Seminars, International Stockman's School**, edited by Frank H. Baker and Mason E. Miller, 12-19. Morillton, AK: Winrock International, 1985.

Outlines the possible uses of genetic technology to improve the beef industry. Emphasizes the role of breeding stock.

773. Wood, Charles L. "Upbreeding Western Range Cattle: Notes on Kansas, 1880-1920." **Journal of the West** 16 (January 1977): 16-28.

Kansas cattlemen contributed to the upbreeding of livestock on the Great Plains by using mixed Spanish and British animals during the 1870s and 1880s. Notes the importance of Shorthorn, Galloway, Aberdeen-Angus and Hereford breeds. Between 1900 and 1920, breeders sought increased meat production and the use of less feed. Some discussion of the work of Robert H. Hazlett with Herefords.

DAIRYING

774. Alvord, Henry E. "Dairy Development in the United States." In **Yearbook of Agriculture 1899**, 381-402. Washington, D.C.: United States Department of Agriculture, 1899.

An overview of American dairying from the colonial period. Economic in emphasis but some useful information about breeding improvements as well as technological change that enabled the development of cheeseries and creameries and the condensed milk industry.

775. Becker, Raymond B. "American Contributors to Better Dairy Cattle." **Hoard's Dairyman** 98 (September 10, 1953): 736-38.

Brief descriptions of the major breeds and their contribution to American dairying, such as Ayrshires, Brown Swiss, Guernseys, Frieslands and Jerseys.

776. _____. **Dairy Cattle Breeds: Origins and Development**. Gainesville: University of Florida Press, 1973. 554 pp.

A history of cattle from the prehistoric age until the twentieth century. Includes breed development with emphasis on the significance of Mendel's laws. Some discussion of disease control and nutrition.

777. Cameron, Jenks. **The Bureau of Dairy Industry: Its History, Activities and Organization**. Baltimore: Johns Hopkins Press, 1929. 74 pp.

An early agency history with some discussion of breeding experiments and engineering developments. Essentially an institutional history with little about its scientific and technological contributions.

778. Conn, H.W. **Practical Dairy Bacteriology**. New York: Orange Judd Co., 1914. 314 pp.

Discusses the practical and theoretical uses of bacteria in relation to the production of milk, cheese and butter as well as the public health. A dated but useful study that suggests directions for bacteriological analysis and experiments.

779. "Milchwirtschaftliche Technologie: Einst und Jetzt." **Deutsche Milchwirtschaft** 38 no. 25 (1987): 863-878.

Traces the development of dairy technology with emphasis on the twentieth century. See for separators, butter and cheese making and cleaning, pumping and cooling technology.

780. Davis, J.G. "Cheesemaking in Britain: The Past and the Future." **Journal of The Society of Dairy Technology** 34 no. 2 (1981): 47-52.

Surveys the origin and development of cheese making in Britain and discusses the contributions of the major cheese makers during the nineteenth century. Contends that cheese making will not become a true technology or applied science until microorganisms can be controlled or eliminated.

781. _____. "Personal Recollections of Developments in Dairy Bacteriology over the Last Fifty Years." **Journal of Applied Bacteriology** 55 no.1 (1983): 1-12.

Reflections on the changes in dairy bacteriology in Great Britain since 1927. Includes brief historical information about the formation of the Society for Applied Bacteriology, Society of Dairy Technology and Society for General Microbiology as well as the work of the Hannah Research Institute and the Commonwealth Bureau of Dairy Science and Technology.

782. Frahm, K. **Rinderrassen in den Landern der Europaischen Gemeinschaft**. Stuttgart: Ferdinand Enke Verlag, 1990. 241 pp.

Describes the origin and domestication of 135 dairy and beef cattle breeds in the European community. Some discussion of breeding history.

783. Frantz, Joe B. **Gail Borden, Dairyman to a Nation**. Norman: University of Oklahoma Press, 1951. 310 pp.

The standard biography that discusses his failures as well as successes, with emphasis on his life in Texas, New York and Connecticut. Discusses Borden's work to keep milk fresh for long periods of time.

784. Fraser, Wilber J. "The Formation of the American Dairy Science Association." **Journal of Dairy Science** 16 (November 1933): 583-586.

Brief history of the organization that formed at the University of Illinois on July 7, 1906. Includes listing of the leaders of the association.

785. Gorner, F. "Der Brinsekase aus Schafmilch (Brumsen)." **Ernahrung** 4 (1980): 157-162.

A discussion of the history, technology and microbiology of making Bryndza cheese in Czechoslovakia. This ewe's milk cheese is made by using lactic acid bacteria and an anaerobic process. The cheese also is known as Liptovska Bryndza.

786. Harris, G. "Brown Swiss: Eine Geschichte von Import und Export." **Tierzuchter** 42 no. 6 (1990): 259-61.

Brief survey of the history of the American Brown Swiss. Notes matters of selection and performance.

787. Hodgson, Ralph E. "Dairy Production Research in the United States Department of Agriculture, 1895 to 1980: A Historical Review." United States Department of Agriculture, **Miscellaneous Publication, No. 1447**, 1986. 57 pp.

Traces the contributions of USDA scientists to breeding and nutrition improvements. Includes some discussion of federal-state cooperative programs as well as a list of the USDA scientists involved with dairy research.

788. Ihde, Aaron J. "Stephen Moulton Babcock--Benevolent Skeptic." In **Perspectives in the History of Science and Technology**, edited by Duane H.D. Rooler, 271-82. Norman: University of Oklahoma Press, 1971.

A brief, favorable discussion of Babcock's life and work. A good place to start.

789. Irwin, P.G. "The Dairying Industry of Kaira, Western India." **South Asia** (1971): 122-131.

A survey of the Kaira District Cooperative Milk Producers Union from 1951 to 1969. The author contends that modern, western dairy technology enabled the farmers to improve their economic gains in Giyarat, West India.

790. Johnson, E.A., J.H. Nelson, and M. Johnson. "Microbiological Safety of Cheese Made from Heat-treated Milk." **Journal of Food Protection** 53 (May 1990): 441-52.

Reviews the history of pasteurization since the late nineteenth century. Notes the success in New Zealand and the controversy in the United States. Describes technological improvements.

791. Kroger, M., J.A. Kurmann, and J.L. Rasic. "Fermented Milks--Past, Present, and Future." **Food Technology** 43 (January 1989): 92, 94-97, 99.

A brief overview of the development of fermented milk products, such as yogurt, cultured buttermilk and sour cream.

792. Lythgoe, H.C. "Historical Survey of Methods for the Determination of the Fat Content of Milk." Association of Food and Drug Officials of the United States, **Quarterly Bulletin** 12 (April 1948): 53-63.

Brief discussion of the methods developed during the late nineteenth century. Notes the work of Ritthausen, Konig, Soxhlet, Wollney, Parsons, Short, Faber and Babcock.

793. McCullough, M.E. "Milestones in Dairy Cattle Feeding." **Hoard's Dairyman** 121 (July 10, 1976): 793, 812.

A note on feeding methods since the colonial period. Feed analysis began in Germany during 1857. After World War II, scientists made great advances in their research on vitamins and minerals to enable feeds to be mixed to suit the needs of a particular cow.

794. Mangurkar, B.R. "Role of Non-Government Agencies in Dairy Cattle Breeding Programme in India." **Silver Jubilee Celebrations, Veterinary College Anand, Lead Papers and Abstracts, National Seminar on Genetics Applied to Livestock**

Production, 23-25 October 1989, A1-A6. Anand, India: Veterinary College, 1989.

A brief history of genetic improvement programs in India. Includes a discussion of dairy cooperatives and organizations in relation to crossbreeding and embryo transfer.

795. Melkonyan, M.S., V.N. Sergeev, and Z. Dilanyan, ed. **Sovremennaya Teknologiya Syrodeliya i Bezotkhodnaya Pererabotka Moloka. Materialy Vsesoyuznoi Nauchno Tehknicheskoi Konferentsii (Erevan, 14-16 Noyabrya 1988**. Erevan, Armenian SSR: Erevanskii Zooveterinarnyi Institut, 1989. 568 pp.

A history of cheese making in the Soviet Union. The papers from this conference have English summaries.

796. Pirtle, T.R. **History of the Dairy Industry**. Chicago: Majonnier Bros. Co., 1926. 646 pp.

A dated but useful history of the science of dairying. Good for comparative purposes.

797. Porter, A.R., J.A. Sims, and C.F. Foreman. **Dairy Cattle in American Agriculture**. Ames: Iowa State University Press, 1965. 328 pp.

Surveys economic, geographical and climatic influences on the development of cattle for dairying and the emergence of distinct breeds. Includes the effects of artificial insemination and improved breeding practices.

798. Prentice, Ezra P. **American Dairy Cattle: Their Past and Future**. New York: Harpers & Brothers Pub., 1942. 453 pp.

Describes the history of Ayrshires, Brown Swiss, Holsteins, Jerseys and Shorthorns from their European origins to breed improvements in the United States.

799. Schlebecker, John T. **A History of American Dairying**. Chicago: Rand-McNally and Co., 1967. 48 pp.

A useful introduction. Discusses matters of disease, such as brucellosis and tuberculosis, as well as technology, such as the Babcock butter fat tester.

800. Trout, George Malcolm. "Homogenized Milk: A Review and Guide." **Memoir Bulletin of the Michigan Agricultural Experiment Station, No. 9**. East Lansing, MI: State College Press, 1950. 232 pp.

A summary of the research on the homogenization of milk at the Michigan Agricultural Experiment Station since 1930. Includes an evaluation of the Babcock, Gerber, Minnesota and Pennsylvania methods and the technology for determining the butter fat content of milk.

801. Westhoff, Dennis C. "Heating Milk for Microbial Destruction: A Historical Outline and Update." **Journal of Food Protection** 41 (February 1978): 122-130.

Dairies first heated milk to improve preservation and later to prevent disease. Discusses thermal processing to sterilize milk for distribution without refrigeration.

802. Wiggans, G.R. "National Genetic Improvement Program for Dairy Cattle in the United States." **Journal of Animal Science** 69 no. 9 (1991): 3853-60.

Brief historical review. Technical matters concerning genetic improvement, but the historian of agricultural science should consult this source for the current state of dairy cattle performance evaluations.

803. Zannoni, M., and L. Zannoni. "Parmesan Cheese: Historical Notes on the Development of Technology." **Scienza e Tecnica Lattiero Casearia** 40 no. 5 (1989): 321-39.

Brief discussion of the history of Parmesan cheese the development of which dates before the thirteenth century. Detailed description of technical developments for production since the early nineteenth century.

SHEEP

804. Bowie, G.G.S. "New Sheep for Old: Changes in Sheep Farming in Hampshire, 1792-1879." **Agricultural History Review** [Great Britain] 35 pt. 1 (1987): 15-24.

A general discussion of sheep raising including breeding improvements in this area of England.

805. _____. "Northern Wolds and Wessex Downlands: Contrasts in Sheep Husbandry and Farming Practice, 1770-1850." **Agricultural History Review** [Great Britain] 38 pt. 2 (1990): 117-26.

Surveys the history of sheep raising, including the introduction of various breeds, the use of manure for fertilizer and pasture rotations.

806. Combs, W. "A History of the Barbados Blackbelly Sheep." In **Hair Sheep of Western Africa and the Americas: A Genetic Resource for the Tropics**, edited by H.A. Fitzhugh and G.E. Bradford, 179-197. Boulder: Westview Press, 1983.

Suggests the Blackbelly originated on Barbados but crossbred with sheep introduced from Africa during the seventeenth century. As a result, a new breed developed well-adapted to tropical environments.

807. Connor, L.G. "A Brief History of the Sheep Industry in the United States." **Annual Report of the American Historical Association, 1918**, 89-197. Vol. 1. Washington, D.C.: Government Printing Office, 1921.

Economic in emphasis, but a useful overview of sheep raising in the United States during the nineteenth and early twentieth centuries. Brief information about breeds and improvements.

808. Copus, A.K. "Changing Markets and the Development of Sheep Breeds in Southern England, 1750-1900." **Agricultural History Review** [Great Britain] 37 pt. 1 (1989): 36-51.

Discusses breeding in relation to prices for mutton, tallow and wool.

809. Harmsworth, T., and G. Day. **Wool and Mohair: Producing Better Natural Fibres**. Victoria: Inkata Press, 1990. 222 pp.

Discusses the production principles applied in Australia to help farmers improve the breeding of sheep for wool and Angora goats for mohair. See also for the history of breeding black and colored sheep.

810. Hickertseder. A. "Parasitenproblems Beim Haustier im Mittelalter und Ihre Wichtigsten Behandlungsverfahren." Inaugural diss., Tierarztliche Fakultat, Ludwig Maximilians Universitat, Munchen, Germany, 1989. 160 pp.

Discusses the problems of parasites for domestic animals during the Middle Ages and methods for relief. Some discussion of sheep and swine, although most of the study deals with animals belonging to the nobility, such as horses and dogs.

811. Kiel, F.W., and M.Y. Khan. "Brucellosis in Saudi Arabia." **Social Science and Medicine** 29 no. 8 (1989): 999-1001.

Discusses brucellosis in relation to current public health in Saudi Arabia. The problem is serious because of the large number of sheep and goat farms, the custom of drinking warm milk from sheep and goats, importations and migrating grazing practices.

812. Konig, K.H., ed. **Tierproduction: Schafzucht**. Berlin: VEB Deutscher Landwirtschaftsverlag, 1990. 194 pp.

A collection of essays on sheep breeding in the German Democratic Republic and around the world. Useful for the agricultural historian.

813. McGregor, Alex. "From Sheep Range to Agribusiness: A Case History of Agricultural Transformation on the Columbia Plateau." **Agricultural History** 54 (January 1980): 11-27.

Discusses the demise of the sheep industry between the nineteenth century and the 1970s. Emphasizes economics, but useful for gaining a historical overview.

814. MacKenzie, A. "Prospects, Problems and Pitfalls in Wool Harvesting." **Agricultural Engineering** [Australia] 17 no. 2 (1988): 1-3.

Historical outline of wool harvesting in Australia. Discusses the technology designed to speed the process and reduce labor costs.

815. McQuirk, B., G.P. Davis, and K. Whiteley. "Sheep Breeding Research in China: The ACIAR Project." **Wool Technology and Sheep Breeding** 36 no. 1 (1988): 28-32.

Surveys the Chinese efforts to improve fine-wooled sheep, which began during the 1950s. Brief information about using Australian Merino genes to improve Chinese sheep.

816. Maijala, K. "History, Recent Development and Uses of Finnsheep." **Journal of Agricultural Science in Finland** 60 no. 6 (1988): 449-54.

Discusses the history of these Finnish sheep in relation to ancestry, fertility and meat and wool production.

817. Masy, Charles. **The Australian Merino**. Ringwood, Australia: Viking O'Neil, 1990. 1,033 pp.

An extensive history of this fine-wooled sheep. See for a general discussion of sheep raising in Australia. Includes an extensive bibliography.

818. Pope, L.S. "Animal Science in the Twentieth Century." **Agricultural History** 54 (January 1980): 64-70.

Since the turn of the twentieth century, scientists have improved sheep breeding, formulated feed supplements and developed new technology that revolutionized sheep raising.

819. Sierra-Alfranca, I. **La Raza Ovina Salz: Creacion y Resultados**. Zaragoza, Spain: IberCaja, 1989. 95 pp.

A detailed history of the Salz sheep, a synthetic breed, since the late 1970s. This breed resulted from the crossing of

Romanov rams with Aragonese ewes. It produces a medium-quality wool.

820. Weir, Harrison. **The Sheep and Pigs of Great Britain: Being a Series of Articles on the Various Breeds of Sheep and Pigs of the United Kingdom, Their History, Management, Etc.**, edited by John Coleman. London: The Field Office, 1877. 214 pp.

Based on the observations and work of Weir (1824-1906). Includes illustrations. Dated but useful.

821. Wentworth, Edward Morris. **America's Sheep Trails: History, Personalities**. Ames: Iowa State College Press, 1948. 667 pp.

Essentially an economic history with little about science and technology, but a useful introduction to the history of sheep raising in the United States.

SWINE

822. Bustad, L.K., and V.G. Horstman. "Pigs: From B.C. to 2000 A.D.: From Outhouse to Penthouse." In **Swine in Biomedical Research**, edited by M.E. Tumbleson, 3-15. New York: Plenum Press, 1986.

A brief history of swine and their use in medical research.

823. Craft, W.A. "Swine Breeding Research at the Regional Swine Breeding Laboratory." United States Department of Agriculture, **Miscellaneous Publication, No. 523**, 1943. 14 pp.

Presents the results of cooperative research at the agricultural experiment stations. Emphasizes the headquarters laboratory at Ames, Iowa. Focuses on breeding, growth rates, weight gains and vitality of swine.

824. Davis, Joseph R., and Harvey S. Duncan. **History of the Poland China Breed of Swine**. Omaha: Poland China History Association, 1921. 278 pp.

A look at the breed from its origins in the mid-1880s to 1920. Little about scientific development.

825. Ensminger, M.E., and R.O. Parker. **Swine Science**. Danville, IL: Interstate Printers & Publishers, 1984. 568 pp.

A general study designed for the producer that includes topics such as breed descriptions, genetics, nutrition and disease.

826. Kolatha, G. "Testing for Trichinosis." **Science** 227 (1985): 621-24.

Discusses the history of trichinosis and new testing procedures.

POULTRY

827. Branion, H.D. "These I Have Known." **Poultry Science** 55 (May 1976): 835-840.

A tribute to the leaders and founders of the Poultry Science Association. Useful biographical material.

828. Crawford, R.D., ed. **Poultry Breeding and Genetics**. Amsterdam: Elsevier, 1990. 1,223 pp.

A comprehensive reference book for the poultry geneticist. Includes the discussion of poultry history and breeding practices.

829. Desrosier, N.W., et al., "What Cathode Rays Do to Eggs and to Milk." **Food Engineering** 27 (May 1955): 78-79, 214.

A note on the benefits of treating eggs and milk with cathode rays to improve storage without the use of refrigeration.

830. Ernst, R.A., J.R. Millam, and F.B. Mather. "Review of Life-History Lighting Programs for Commercial Laying Fowls." **World's Poultry Science Journal** 43 no. 1 (1987): 45-55.

Discusses lighting in relation to feeding, laying and genetic differences. See for any study dealing with the technological change of the poultry industry.

831. Hanke, Oscar A., and John L. Skinner, ed. **American Poultry History, 1823-1973**. Madison: American Poultry Historical Society, 1974. 775 pp.

Essentially an economic and institutional history. Some discussion of science and technology in relation to breeding, disease, nutrition and hatcheries. No critical analysis.

832. Hewson, P. "Origin and Development of the British Poultry Industry: The First Hundred Years." **British Poultry Science** 27 no. 4 (1986): 525-539.

Dates the British poultry industry to the 1890s with the establishment of the National Institute of Poultry Husbandry at Harper Adams College. Scientists began to investigate the practical applications of genetics to poultry breeding. Discusses the influence of controlled environments after 1945 on poultry diseases.

833. Jull, M.A., et al. "The Poultry Industry." In **Yearbook of Agriculture, 1924**, 377-456. Washington, D.C.: United States Department of Agriculture, 1924.

A historical survey with some discussion of breeding and disease, but essentially economic in focus.

834. Lake, P.E. "The History and Future of the Cryopreservation of Avian Germ Plasm." **Poultry Science** 65 (January 1986): 1-15.

Traces the history of freezing cellular material, such as germ tissue, embryos and eggs. Emphasizes the current status of cryopreservation and suggests the possibilities of this science for preserving germplasm for future breeding experiments.

835. Moreng, R.F. "Evaluating Incubation Practices." **Proceedings of the Australian Poultry Science Symposium**, 62-68, 1990.

Surveys the history of artificial incubation.

836. Orozco, Pinan F. **Razas de Gallinas Espanolas**. Madrid: Ediciones Mundia Prensa, 1989. 216 pp.

A detailed history of Spanish poultry breeds.

837. Skinner, J.L. "The Beginnings and Development of Poultry in North America." In **Exploring Our Livestock Heritage: Proceedings of the 1988 Annual Meeting of the Minor Breeds Conservancy**, 50-57. Pittsboro, NC: The Conservancy, 1988.

Brief description of the early history of poultry breeds and egg and meat production.

838. Sykes, A.H. "An Introduction to the History of Incubators." In **Avian Incubation**, edited by S.G. Tullett, 297-303. Kent, United Kingdom: Butterworths, 1990.

Describes the technology and use of incubators from 1750 to 1950.

839. Young, R.J. "Poultry Nutrition: A Twentieth Century Achievement." **Cornell Veterinarian** 75 (January 1985): 230-247.

Discusses nutrition research regarding energy, protein, vitamins, amino and fatty acids and inorganic elements. Notes the relationship of each to producing scientifically developed feed formulas that are the most cost efficient for the production of meat and eggs.

CHAPTER XII

ENTOMOLOGY

840. Allen, Henry W. "The History of the American Entomological Society." **Transactions of the American Entomological Society** 85 (December 1959): 335-372.

    A brief history of the society founded in 1859, with emphasis on biographical sketches of prominent members.

841. Allen, W.A., and E.G. Rajotte. "The Changing Role of Extension Entomology in the IPM Era." **Annual Review of Entomology** 35 (1990): 379-97.

    A brief history of Integrated Pest Management (IPM) in the extension system since 1972. Based on interviews with extension agents, some comparison of technology.

842. Barnes, J.K. "Insects in the New Nation: A Cultural Context for the Emergence of American Entomology." **Bulletin of the**

Entomological Society of America 31 (Spring 1985): 21-30.

Discusses the social, political and economic developments that shaped the emergence of entomology in the United States. Reviews the influence of agricultural science and societies.

843. Basden, E.B. "James Hardy's Entomological Notebooks." **Journal of the Society for the Bibliography of Natural History** 4 (1962-1968): 44-47.

A topical list of Hardy's notebooks held by the Berwickshire Naturalists' Club at the Museum and Art Gallery, Berwick on Tweed, England. Hardy (1815-1898), a Berwickshire naturalists, had a strong interest in entomology.

844. Carpenter, M.M. "Bibliographies of Biographies of Entomologists." **American Midland Naturalist** 33 (January 1945): 1-116.

Updates the bibliography of biographies of entomologists published by J.S. Wade in the **Annals of the Entomological Society of America** 21 (1928). International in scope and complete through 1943.

845. Casagrande, R.A. "The Colorado Potato Beetle: 125 Years of Mismanagement." **Bulletin of the Entomological Society** 33 (Fall 1987): 142-150.

A review of the potato beetle's migration from Mexico and the methods used to combat this insect. Includes the use of pesticides and the beetle's resistance to chemical controls.

846. Caltagirone, L.E., and R.L. Doutt. "The History of the Vedalia Beetle Importation to California and Its Impact on the Development of Biological Control." **Annual Review of Entomology** 34 (1988): 1-16.

A good introduction to the history of the lady bug in California agriculture since the late nineteenth century. Shows the importance and the effectiveness of biological control under certain circumstances.

847. Choi, K.M., and M.H. Lee. "The Use of Natural Enemies for Controlling Agricultural Pests in Korea." In **FFCT-NARC International Seminar on "The Use of Parasitoids and Predators to Control Agricultural Pests" Tukuha Science City, Ibaraki-ken, 305 Japan, October 2-7, 1989**. Tukuba-gun, Japan: National Agricultural Research Centre, 1990. 20 pp.

Generally, Korean farmers have used chemicals to eradicate insects, and their history of natural controls is brief. Korean scientists, however, are giving increased attention to biological controls, and they have identified 502 species capable of helping farmers combat unwanted insects.

848. Cloudsley-Thompson, J.L. **Insects and History**. London: Weidenfeld & Nicholson, 1976. 242 pp.

Although this study does not discuss agricultural entomology, it provides an important overview of the history of insects and disease since ancient times. Agricultural scientists will find it useful for background information, context and synthesis.

849. Cortes, R., and J. Herrera. "Historical and Biological Background for a History of Entomology in Chile." **Acta Entomologica Chilena** 15 (1989): 297-322.

An introduction to the history and historiography of entomology in Chile. A good source for beginning work in this area.

850. Cresson, E.T. **A History of the American Entomological Society**. Philadelphia, 1909. 60 pp.

A dated history, but still helpful when used with more contemporary histories.

851. Dodge, Charles Richards. "The Life and Entomological Work of the Late Townend Glover, First Entomologist of the United States Department of Agriculture." United States Department of Agriculture, Division of Entomology, **Bulletin 18**, 1888. 68 pp.

A dated but useful biography that includes a bibliography.

852. Dunlap, Thomas R. "Farmers, Scientists and Insects." **Agricultural History** 54 (January 1980): 93-107.

During the first two decades of the twentieth century, the Bureau of Entomology in the USDA grew rapidly. An emphasis on commercial agriculture, major insect threats and a faith in science increased the importance of the bureau.

853. Farber, P.L. "A Historical Perspective on the Impact of the Type Concept on Insect Systematics." **Annual Review of Entomology** 23 (1978): 91-99.

Examines the origin and use of the word "type" in entomology before 1859 to show the complexity of that concept and to evaluate its role in the biological sciences prior to Darwinism.

854. Flent, Oliver S. **The United States National Entomological Collections**. Washington: Smithsonian Institution Press, 1976. 47 pp.

Surveys the research collections at the National Museum of Natural History in Washington, D.C.

855. Fuester, R.W. "The Cressons and The American Entomological Society." **Entomological Review** 95 (September/October 1984): 149-154.

A brief discussion of the contributions of Ezra T. Cresson, Sr. (1838-1925) and his two sons, George B. (1859-1919) and Ezra T., Jr. (1876-1948) to the science of entomology and the American Entomological Society.

856. Gilbert, P., and C.J. Hamilton. **Entomology: A Guide to Information Sources**. 2nd ed. London: Mansell Publishing, 1990. 259 pp.

A worldwide guide to the study of insects. Includes a survey of the history and early literature.

857. Gupta, A.P., ed. **Resources in Entomology**. College Park, MD: Entomological Society of America, 1987. 269 pp.

Contains listings of the entomological resources in Western Europe, Australia, Africa, Latin America and the United States. Useful for information about institutions, organizations and scientists.

858. Goldman, Abe. "Tradition and Change in Postharvest Pest Management in Kenya." **Agriculture and Human Values** 8 (Winter-Spring 1991): 99-113.

Discusses the farmer's response to postharvest crop losses by region and tradition. Notes the influence of population growth and new crops on changing techniques. A social and cultural focus rather than scientific or technological.

859. Hallman, G.J., A.R. Panizzi, and W.H. Whitcomb. "Entomological Societies in Latin America." **American Entomologist** 38 no. 1 (1992): 22-32.

A survey of institutional development.

860. Hawes, Ina L., and Rose Eisenberg. "Bibliography of Aviation and Economic Entomology." United States Department of Agriculture, **Bibliographical Bulletin No. 8**, 1947. 186 pp.

Covers the years 1919 through 1944. International in scope. References relate to the use of aircraft to control crop pests, forest insects and mosquitos.

861. Hatch, Melville H. **A Century of Entomology in the Pacific Northwest**. Seattle: University of Washington Press, 1949. 42 pp.

Traces the significant events, identifies the major scientists and discusses the most important books, papers and collections relating to the development of entomology in this region. Includes an examination of the status of entomology in the Pacific Northwest of the United States.

862. Helms, Douglas. "Revision and Reversion: Changing Cultural Control Practices for the Cotton Boll Weevil." **Agricultural History** 54 (January 1980): 108-125.

An excellent study of the methods used, such as plowing and burning, to eradicate the boll weevil prior to the development of chemical insecticides. Includes an evaluation of the contributions of Seaman A. Knapp and Leland O. Howard. Helms is the authority for the history of the boll weevil in the United States.

863. _____. "Technological Methods for Boll Weevil Control." **Agricultural History** 53 (January 1979): 286-99.

Discusses the methods that farmers and scientists have used to control the boll weevil, such as poisoning, trapping and planting early varieties of cotton. Pesticides, however, proved the most effective, but technological control methods were often expensive and contributed to the consolidation of farms in the South.

864. Holland, W.J. "The Development of Entomology in North America." **Annals of the Entomological Society of America** 13 (1920): 1-15.

Discusses the insect studies made by Europeans on the North American continent during the eighteenth and nineteenth centuries. Notes the first paper presented on entomology in the American colonies by John Bartram in 1745. Discusses the increasing interest in the study of insects that affect agriculture.

865. Howard, Leland Ossian. "A History of Applied Entomology." Smithsonian Institution, **Miscellaneous Collections No. 84**, 1930. 564 pp.

Examines the development of entomology worldwide. Notes the most important people, governmental activities, professional organizations and scientific journals. Good for a comparative, international overview.

866. _____. "Progress in Economic Entomology in the United States." In **Yearbook of Agriculture 1899**, 135-56. Washington D.C.: United States Department of Agriculture, 1899.

     Discusses the work of William D. Peck, Thaddeus W. Harris and Asa Fitch and the work of the federal government to fund studies in economic entomology during the mid-nineteenth century. Some information on international developments.

867. _____. "The Rise of Applied Entomology in the United States." **Agricultural History** 3 (July 1929): 131-139.

     Discusses the major entomologists in the United States and the actions of Congress related to entomology. Begins with the work of William D. Peck in 1795, examines the importance of the Hatch Act in 1887 and highlights contemporary achievements.

868. "Howard, Leland Ossian, 1857-1950." **Journal of Economic Entomology** 43 (December 1950): 958-962.

     An obituary of one of the world's most important entomologists. Emphasizes his professional career.

869. Klassen, W., and P.H. Schwartz Jr. "ARS Research Program in Chemical Insect Control." In **Agricultural Chemicals of the Future: Invited Papers Presented at a Symposium Held May 16-19, 1983, at the Beltsville Agricultural Research Center**,

edited by James L. Hilton, 267-291. Totowa, NJ: Rowman & Allanheld, 1985.

In 1881, the United States Department of Agriculture began a research program for the chemical control of insects. Recent work in the department's Agricultural Research Service has stressed the evaluation, behavior and fate of synthetic insecticides.

870. Lofgren, C.S. "History of Imported Fire Ants in the United States." In **Fire Ants and Leaf-Cutting Ants: Biology and Management**, edited by Clifford S. Lofgren and Robert K. Vander Meer, 36-47. Boulder: Westview Press, 1986.

Traces the history of the fire ant in the United States from the 1930s to the mid-1980s. Notes control and eradication methods of the USDA and the problems that developed with the use of pesticides.

871. Mallis, Arnold. **American Entomologists**. New Brunswick: Rutgers University Press, 1971. 549 pp.

Discuss the major accomplishments of North American entomologists but emphasizes their personal, rather than scientific lives, whenever possible.

872. Ordish, George. **The Constant Pest: A Short History of Pests and Their Control**. New York: Charles Scribner's Sons, 1976. 240 pp.

A brief history of pests, especially insects from the neolithic age to the twentieth century. Good overview of pest control methods. Designed for the general reader or introductory student.

873. Osborn, Herbert T. **Brief History of Entomology**. Columbus, OH: Spahr & Glenn, 1952. 303 pp.

Discusses the development of entomology and the scientists involved. Includes information about the most important colleges, publications, societies and collections for this profession.

874. Parencia, C.R. "One Hundred Twenty Years of Research on Cotton Insects in the United States." United States Department of Agriculture, Agricultural Research Service, **Agricultural Handbook No. 515**, 1978. 75 pp.

Summarizes the research on cotton insects and their control by the Agricultural Research Service and the agricultural experiment stations. Summarizes the results of early and current research on control methods.

875. Perkins, John H. "Edward Fred Knipling's Sterile-Male Technique for Control of the Screw-worm Fly." **Environmental Review** 5 (1978): 19-37.

Between 1937 and 1955, Knipling developed the complex method of making the male fly infertile to help reduce the population of this insect which damages the cattle industry in the southern United States. This study shows that a unique combination of biology and culture was necessary to perfect this technique.

876. _____. "Insects, Food, and Hunger: The Paradox of Plenty For U.S. Entomology, 1920-1970." **Environmental Review** 7 no. 1 (1983): 71-96.

Explores the relationships between invention and innovation in relation to the control of pests, the food supply and hunger. Focuses on the use of pesticides.

877. _____. "Reshaping Technology in Wartime: The Effect of Military Goals on Entomological Research and Insect-Control Practices." **Technology and Culture** 19 (April 1978): 169-86.

Although DDT was developed prior to 1939, World War II accelerated its use. After the war, DDT and other synthetic pesticides disrupted biological and cultural control practices at the expense of the environment.

878. Pimentel, David, ed. **Insects, Science, and Society--Proceedings of a Symposium, Ithaca, New York, October, 1974**. New York: Academic Press, 1975. 284 pp.

Discusses recent research in entomology and the practical problems of pest control. Good historical perspective on economic entomology, including its social implications.

879. Riegert, Paul W. **From Arsenic to DDT: A History of Entomology in Western Canada**. Toronto: University of Toronto Press, 1980. 357 pp.

Encyclopedic in its coverage on agricultural entomology as well as government support for research. Emphasizes early collectors and naturalists, professional entomologists and their societies to 1914 and British Columbia and the prairie provinces.

880. Roback, S.S. "James A.G. Rehn and the American Entomological Society." **Entomological News** 95 (September/October 1984): 163-65.

Biographical sketch of Rehn's work as the editor of the **Transactions** from 1917-1924 and his service to the Society.

881. Russell, L.M. ."Leland Ossian Howard: A Historical Review." **Annual Review of Entomology** 23 (1978): 1-15.

Examines the career of Howard as an economic entomologist before that field had professional recognition. Howard served as an entomologist in the USDA from 1878 to 1931, and he directed the entomological work of the department from 1894 to 1927. He helped make entomology a legitimate branch of the biological sciences.

882. Sanborn, C.E. "History and Control of the Boll Weevil in Oklahoma." Oklahoma Agricultural Experiment Station, **Bulletin 222**, 1934. 32 pp.

Describes the chemical methods used to control the boll weevil, especially calcium arsenate. Includes a discussion of the history of the boll weevil which first struck Oklahoma in 1905, when cotton was the major cash crop.

883. Smith, Edward H. "The Comstocks and Cornell: In the People's Service." **Annual Review of Entomology** 21 (1976): 1-25.

Discusses the lives of two important entomologists at Cornell University during the late nineteenth century. Reviews their published research and examines Anna Comstock's role as a professional entomologist.

884. Smith, Ray F., Thomas E. Mittler, and Carroll N. Smith, ed. **History of Entomology**. Palo Alto, CA: Annual Reviews, 1973. 517 pp.

Traces the relationship between insects and culture from the ancient world to the present. Discusses the ways that man has attempted to use insects and to minimize their damage to crops. This study also traces the general development of insect morphology, physiology and ecology. Notes recent developments in economic entomology and includes considerable biographical information.

885. Sorensen, Willis Conner. **Brethren of the Net: American Entomology, 1840-1880**. Ph.D. diss., University of California, Davis, 1984. 432 pp.

Between 1840 and 1880, American scientists became leaders in the field of entomology. During that period they organized professionally and provided proof that their science would help improve agricultural production and profits.

886. _____. "The Rise of Government-Sponsored Applied Entomology, 1840-1870." **Agricultural History** 62 (Spring 1988): 98-115.

During the mid-nineteenth century, public support provided the stimulus for the development of agricultural entomology as an agricultural science. As farmers attempted to increase production and profits, they called on entomologists to solve their problems. State and federal entomologists soon were appointed and entomological teaching and research began at agricultural colleges. Although entomologists used biological as well as chemical controls, they sought a balance of nature rather than the total destruction of an insect species.

887. Southwood, T.R.E. "Entomology and Mankind." **Proceedings of the International Congress of Entomology**, 36-51. College Park, MD: Entomological Society of America, 1977.

Discusses the relationship between entomology, theology and art. Philosophically based.

888. Spencer, G.J. "A Century of Entomology in Canada." **Canadian Entomologist** 96 (September 1963): 33-59.

Biographical sketches of the major scientists in Canadian entomology, such as C.J.S. Bethune, James Fletcher, Gordon Hewitt, Arthur Gibson, Leonard McLaine, H.S. Crawford, Robert Glen and B.N. Smallman.

889. "Fighting Our Insect Enemies--Achievements of Professional Entomology, 1854-1954." United States Department of Agriculture, **Agricultural Information Bulletin, No. 121**. 1954. 23 pp.

A brief history. Noncritical but a good place to begin.

890. Wade, J.S. "A Bibliography of Biographies of Entomologists, With Special Reference to North American Workers." **Annals of the Entomological Society of America** 21 (1928): 489-520.

Emphasizes the scientists who specialized in entomology. A compilation of the most readily available sources, many of which were written by non-American scholars. Dated but useful.

891. Weber, Gustavus Adolphus. **The Bureau of Entomology: Its History, Activities and Organization**. Washington, D.C.: Brookings Institution, 1930. 175 pp.

An detailed study of the Bureau. See for regulatory activities.

892. Wheeler, A.G. "The Tarnished Plant Bug: Cause of Potato Rot?--An Episode In Mid-Nineteenth Century Entomology and Plant Pathology." **Journal of the History of Biology** 14 no. 2 (1981): 317-338.

During the nineteenth century, Alexander Henderson argued that a subterranean insect, rather than a fungus, caused potato blight. Critical and systematic experimentation did not contribute to the scientific knowledge about potato blight and other matters of entomology until the early twentieth century.

CHAPTER XIII

VETERINARY MEDICINE

893. "A Historical Survey of Animal Disease Morbidity and Mortality Reporting: A Report." **National Research Council Publication 1346**. Washington: National Academy of Sciences-National Research Council, 1966. 40 pp.

Although dated, this publication provides useful information for the historian on the collection of American animal morbidity and mortality statistics. Includes a report on disease nomenclature as well as a summary of animal disease reports.

894. **A History of the Oversea Veterinary Services**. 2 Vols. London: British Veterinary Association, 1973.

A review of the British overseas veterinary service, organized by country, sponsored by the British Veterinary Association. Covers Asia, Africa, the Middle East and the British Virgin Islands. Each country summary provides brief information about the veterinary services available from the late nineteenth century to the 1960s.

895. Adamson, P.B. "Diseases Associated With Man and Pigs in the Ancient Near East." **Historia Medicinae Veterinariae** [Denmark] 14 no. 4 (1989): 104-10.

Discusses the diseases of humans that might have affected swine. Describes bacterial, viral and protozoal infections.

896. Alexander, D.J. "Historical Aspects." In **Newcastle Disease**, edited by D.J. Alexander, 1-10. Boston: Kluwer Academic Publishers, 1988.

This brief introduction provides a good place to start.

897. **Animal Health: Livestock and Pets, Yearbook of Agriculture, 1984**. Washington, D.C.: Government Printing Office, 1984. 646 pp.

An up-to-date, easy-to-use handbook on animal health and disease. Due to the constant improvements in the care of livestock, historians will also want to consult the 1942 yearbook, **Keeping Livestock Healthy**, as well as the 1956 yearbook, **Animal Diseases**. As early as 1899, these yearbooks provided overviews of livestock diseases in the United States.

898. Auty, J.H. "Veterinary Science in Australia: The Colonial Period." **Historia Medicinae Veterinariae** [Denmark] 8 no. 2 (1983): 34-36.

A short article on the early years of the Australian veterinary profession. Qualified veterinarians were immigrating as early as 1840. In 1880, William Tyson Kendall arrived, and he was able to organize a regulatory board and create a college by 1888.

899. Bierer, Bert W. **History of Animal Plagues of North America With an Occasional Reference to Other Diseases and Diseased Conditions**. Washington, D.C.: United States Department of Agriculture, 1974, 97 pp.

A reprint of the 1939 edition, provides a fascinating account, in chronological order, of animal plagues. Useful for students of veterinary history as well as social historians.

900. _____. **A Short History of Veterinary Medicine in America**. East Lansing, MI: Michigan State University Press, 1955. 113 pp.

A standard history for the student of agricultural history or a new veterinarian. This work provides a look at the evolution of the profession from an "indigenous art" to " practitioner" to "trained professionals" that resulted from changes in agriculture, science and education.

901. Braz, M.B. "Contribuicao das Ciencias Veterinarias para a Historia e Desenvolvimento da Ciencia em Portugal." **Revista Portuguesa de Ciencias Veterinarias** 85 (1990): 493-494.

   Notes the relationship of veterinary medicine to the history and development of science in Portugal.

902. Briggs, P.M. "Vaccines and Vaccination: Past, Present and Future." **British Poultry Science** 31 no. 1 (1990): 3-22.

   Reviews the history of poultry diseases and the immunizations and vaccines used to prevent them. Some discussion of biotechnology and recombinant vaccines.

903. Bullis, Kenneth L. "The History of Avian Medicine in the United States: II. Pullorum Disease and Fowl Typhoid." **Avian Diseases** 21 (July-September 1977): 422-29.

   In 1899, Leo F. Retteger discovered Pullorum disease in chickens. Avian pathology became a by-product of his work. Although many cures were attempted from vaccination to the removal of infected birds from the flock, no one method provided better results than the others.

904. _____. "The History of Avian Medicine in the United States: III Salmonellosis." **Avian Diseases** 21 (July-September, 1977): 430-35.

   First reported in pigeons in 1895, this disease had infected ducks by 1920 and turkeys by 1933. By 1960, scientists had identified sixty types of salmonellae. Includes a discussion of its transmission to humans from eating infected poultry.

905. Bundza, A., and Oplustil, P. "A Brief History of Veterinary Medicine, and Its Development in Slovakia." **Historia Medicinae Veterinariae** [Denmark] 9 no. 1 (1984): 1-11.

History of Slovakian veterinary medicine, including treatment of animals and veterinary education during the Austro-Hungarian Monarchy (1867-1918). Brief information on two outstanding veterinarians F. Hutyra and J. Marek.

906. Bustad, Leo K. "An Historic Perspective of Veterinary Medical Education." **Journal of Veterinary Medical Education** 14 no.2 (Fall 1987): 38-43.

A concise evaluation of veterinary education, primarily U.S. and Canadian. Includes a summary of successes and failures.

907. Cohen, Bennett J., and Franklin M. Loew. "Laboratory Animal Medicine: Historical Perspectives." In **Laboratory Animal Medicine**, edited by James G. Fox, Bennett J. Cohen and Franklin M. Loew, 1-17. Orlando, FL: Academic Press Inc., 1984.

This brief history of laboratory animal medicine includes biographical sketches of early veterinarians specializing in laboratory animal medicine. Describes the historical development of the professional associations, including the National Society of Medical Research (NSMR) and the American Association for Laboratory Animal Science (AALAS). Provides information on early training programs.

908. Cole, Clarence G. "History of Hog Cholera Research in the U.S. Department of Agriculture, 1884-1960." United States Department of Agriculture, Agricultural Research Service, **Agriculture Information Bulletin No. 241**, 1962. 124 pp.

An important source for the historian. This bulletin provides a synthesis from the many publications of the Bureau of Animal Industry (1884-1953). Emphasizes the discovery of the virus and the work to provide an immunization. An addendum covers the work of the Animal Disease and Parasite Research Branch (1954-1960). Includes a bibliography as well as a list of patents covering various vaccines.

909. Coubrough, Rhoderick I. "The South African Veterinary Association: Glimpses from 75 Years of Service to Veterinary Science." **Journal of the South African Veterinary Association** 49 no. 4 (December 1978): 283-86.

A short history of the South African Veterinary Association since 1903. Professional education was available by 1920.

910. Crawford, Lester M. "A Tribute to Alexandre Liautard, the Father of the American Veterinary Profession." **Journal of the American Veterinary Medicine Association** 169 (July 1, 1976): 35-37.

A bicentennial tribute to a Frenchman who was responsible for founding the United States Veterinary Medical Association, later to become the American Veterinary Medical Association. In addition, Liautard was a founder of the **American Veterinary Review**, later to become the **Journal of the American Veterinary Medical Association**. He emigrated to the United States in the late 1850's and, by 1859, practiced veterinary medicine in New York City.

911. Diamant, Gerald. "Regulatory Veterinary Medicine: And They Blew a Horn in Judea." **Journal of the American Veterinary Medical Association** 172 (January 1978): 45-54.

History of the various diseases facing American livestock and the eradication programs recommended by the United States Animal Health Association (USAHA) to the USDA. In addition, the article discusses the organization and re-organization of the Animal and Plant Health Inspection Service (APHIS), which is responsible for American regulatory and control programs.

912. Dincer, Ferruh. "Turkiye'de Askeri Veteriner Hekimlik Tarihi Uzerinde Arastirmalar." **Veteriner Fakultesi**

Dergisi Ankara Universitesi 26 no. 3-4 (1979): 1-13.

Discusses veterinary science in the Turkish military. Although training was available as early as 1842 for military veterinarians, a "civilia" veterinary school was not established until 1889.

913. Dolman, Claude E. "Texas Cattle Fever: A Commemorative Tribute to Theobald Smith." **Clio Medica** 4 (March 1969): 1-31.

A tribute to Theobald Smith for his investigations on Texas fever. Smith conducted his research between 1889 and 1892, and published his report in 1893. Dolman argues that Smith did not receive proper credit for his contributions due to professional jealousies.

914. Doughterty, R.M. "A Historical Review of Avian Retrovirus Research." In **Avian Leukosis**, edited by G.F. De Boer, 1-27. Boston: Martinus Nyhoff Pub., 1987.

This study begins in 1908 with the clinical transmission of chicken leukemia. Highly technical discussion of research efforts through 1984.

915. Driesch, Angela von den. **200 Jahre Tierarztlich Lehre und Forschung im Munchen**. Stuttgart: Schattauer, 1990. 212 pp.

Two hundred years of veterinary teaching and research at the Universitat Munchen, with summaries in English.

916. _____. **Geschichte der Tiermedizin: 5000 Jahre Tierheilkunde**. Munchen: Callwey, 1989. 295 pp.

Illustrated survey of the development of veterinary medicine from classical times through the twentieth century.

917. Erasmus, B.J. "The History of Bluetongue." In **Bluetongue and Related Orbiviruses: Proceedings of an International Symposium Held at the Asilomar Conference Center, Monterey, California, January 16-20, 1984**, edited by T. Lynwood Barber, Michael M. Jochim and Bennie I. Osburn, 7-12. New York: A.R. Liss, 1985.

A brief history of research activities on bluetongue. Includes a useful geographic section on outbreaks around the world. Although this is a disease of sheep, other ruminants, such as goats, are susceptible.

918. Fox, Francis H. "Large Animal Medicine: Many Changes in Forty Years." **Cornell Veterinarian** 75 (January 1985): 111-17.

A short history of the diseases encountered and the drugs used in large animal practices. Some of the positive changes have been the development of sulfonamides and antibiotics, the eradication of hog cholera and the control of brucellosis (Bang's disease).

919. Fox, James G. "Laboratory Animal Medicine: Changes and Challenges." **Cornell Veterinarian** 75 (January 1985): 159-70.

Simon D. Brimhall (1863-1941), who worked at the Mayo Clinic from 1915-1922, became the first veterinarian formally employed in laboratory animal medicine. In 1957, the AVMA recognized this speciality of laboratory animal medicine.

920. Giese, Christian. **Die Entwicklung der Tierheilkunde an der Universitat Giessen von den Anfangen bis zum Jahre 1886**. Giessen: W. Schmitz, 1985. 260 pp.

The development of veterinary science at the University of Giessen (Federal Republic of Germany) from its beginning to the year 1866.

921. Gorham, John R. "Biotechnology and Veterinary Medicine." **Proceedings of the United States Animal Health Association** 91 (1987): xxxvi-lxiii.

This paper addresses the influence of biotechnology on veterinary medicine. Includes a literature review. This is a starting place for historians of veterinary science and technology as they begin to examine a new era of genetically engineered livestock. Publication available from the USDA ARS as **Reprint 241** (1988).

922. Habel, Robert E. "Anatomy: Past, Present, Future." **Cornell Veterinarian** 75 (January 1985): 27-55.

An international look at vertebrate anatomy from the ancient Greeks to the present.

923. Harvey, Colin E. "History of Veterinary Dentistry." **Veterinary History: Bulletin of the Veterinary History Society** New Series 5 (Summer 1988): 97-101.

Discusses progress made in techniques in Europe as well as recent changes in American and British veterinary dental practice. This speciality is now incorporated into the veterinary curricula and special societies are being formed.

924. Henderson, William M. "A History of the ARC's Support of Veterinary Research." **Veterinary Record** [Great Britain] 109 (July 11, 1981): 29-34.

A brief history of the Agricultural Research Council (1931-1981). The council has strengthened its reputation for research because the British universities were not prepared or interested in veterinary work. Since 1937, the Council has established a number of institutes to deal with specific issues in animal research.

925. Hosgood, Giselle. "The History of Surgical Drainage." **Journal of the American Veterinary Medical Association** 196 (January 1, 1990): 42-44.

A brief history from the Middle Ages to the present about open drainage techniques. References provided to surgical drainage techniques.

926. Hurt, R. Douglas. "From Superstition to Science: Veterinary Medicine." **Timeline** 1987 4 (June/July 1987): 32-45.

Discusses the development of veterinary medicine in the United States from traditional and often inhumane folk practices for treating disease, sickness and injuries of livestock to the establishment of a scientifically trained profession.

927. Ignace, Jean Alain Ean. **Two Centuries of Veterinary Profession in Mauritius, 1771-1979**. Rose Hill, Mauritius: J.A.E. Ignace, 1984. 133 pp.

An illustrated history of the veterinary profession with emphasis on the contributions of J.D. Shuja, who served in the Veterinary Services from 1956 until 1979.

928. Jakubowski, Stefan. "Nieco Danych o Historii i Dzialalnosci Polskiego Towarzystwa Nauk Weterynaryjnych." **Medycyna Weterynaryjna** 28 no. 2 (Luty 1972): 66-69.

A compilation of facts about the history of the Polish Society for Veterinary Science.

929. Janiszewski, Josef. "Perenc as Historian of Veterinary Science in Poland." **Historia Medicinae Veterinariae** [Denmark] 11 no. 2 (1986): 78-79.

A brief biographical sketch of Aleksander Perenc (1888-1958), a Polish military veterinary surgeon as well as a historian of comparative veterinary medicine.

930. Kalupov, I. "A Century of Veterinary Science in Bulgaria." **Veterinarno Meditsinski Nauki** 19 no. 7 (1982): 6-13.

A brief, uncritical review covering the period 1881-1981.

931. Kapanadze, K.S. "Sbornik Trudov." **Gruzinskii Zootekhnichesko-Veterinarnyi Institut** 38 (1973): 203-213.

Discusses the role of veterinary surgeons in the development of medical science and practice in Russia.

932. Karasszon, Denes. **A Concise History of Veterinary Medicine**. Budapest: Akademiai Kiado, 1988. 458 pp.

An extremely useful overview of the entire history of veterinary medicine. Topically arranged.

933. Katsuyama, Osamu. "The Researchers of the History of Veterinary Medicine in Japan." **Historia Medicinae Veterinariae** [Denmark] 1 no. 5 (1976): 134-136.

A brief look at Japanese veterinary medicine and animal husbandry. Includes a discussion of the Japanese Association for the History of Veterinary Medicine since 1972 and notes the influence of the **Japanese Journal of Veterinary History**.

934. Kester, Wayne O. "Development of Equine Veterinary Medicine in the United States." **Journal of the American Veterinary Medicine Association** 169 (July 1, 1976): 50-55.

A short history of equine veterinary medicine looking at three eras in America: the period prior to World War I; the period between the world wars when tractors and trucks began replacing horses; and the post-war years with the revival of raising horses for pleasure and racing.

935. Kingrey, Burnell W. "Farm Animal Practice in the United States." **Journal of the American Veterinary Medicine Association** 169 (July 1, 1976): 56-60.

A short history of farm animal practice covering epidemics, the ethics of practitioners, the evolution of drugs for treatment, the development of vaccines and the geographical shifts in livestock populations.

936. Knusel-Juvalta, Franz. **Das Zivile Veterinarwesen in der Zentralschweiz von den Angangen bis 1885 und die Geschichte der Gesellschaft Zentralschweizerischer Tierarzte, 1885-1985**. Malters, Switzerland: Die Gesellschaft, 1985. 223 pp.

A history of civil veterinary medicine in central Switzerland until 1885 and the history of the Society of Central Swiss Veterinarians, 1885-1985.

937. Koromyslov, Georgy. "Veterinary Science in the USSR: Current State and Prospects." **USSR Policy Paper No. 90-UPP2**. Ames, Iowa: Center for Agricultural and Rural Development (CARD), Iowa State University, 1990. 11 pp.

    A brief paper updating the status of veterinary medicine as historians look at "perestroika." No references.

938. Kovacs, F., ed. "200 Years of Veterinary Education in Hungary." **Acta Veterinaria Hungarica** 35 nos. 1-2 (1987): 1-209.

    A detailed account of the veterinary research conducted at the University of Veterinary Science. Prior to 1962, this institution was the Veterinary Institute at the Budapest Medical Faculty.

939. Kovalenko, Ia R. "Veterinarnaya Nauka k 50-letiyu Obrazovaniya SSSR." **Trudy Vsesoiuznogo Instituta Eksperimental'noi Veterinarii** 41 (1973): 3-18.

    A brief review of veterinary science in the Soviet Union from 1917 to 1957.

940. Kuttler, Kenneth L. "World-Wide Impact of Babesiosis." In **Babesiosis of Domestic Animals and Man**, edited by Miodrag Ristic, 1-22. Raton, FL: CRC Press, 1988.

    An useful international overview of the history of babesiosis (Texas fever). The author covers the discovery of various babesia and their distribution. Emphasis is on bovine babesiosis, however, brief historical information is provided for horses, sheep and goats. Extensive bibliography.

941. Lancaster, John E., and Julius Fabricant. "The History of Avian Medicine in the United States. IX. Events in the History of Avian Mycoplasmosis, 1905-70." **Avian Diseases** 32 (October/December 1988): 607-23.

A brief history of respiratory diseases of poultry. The authors include an extensive bibliography. This article is part of a series of articles on avian medicine published since 1976 in **Avian Diseases**.

942. Lund, E.E. "The History of Avian Medicine in the United States." **Avian Diseases** 21 (October-December 1977): 459-80.

Discusses the major contributions of American scientists to the study of poultry parasites during the past century. These scientists especially worked to understand diseases attributable to helminths and protozoa in turkeys and chickens respectively.

943. Miller, Everett B. "Comparative Medicine: American Experience From Equine Tetanus--From Benjamin Rush to Toxoid." **Bulletin of the History of Medicine** 57 no. 1 (1983): 81-92.

Although Benjamin Rush first reported tetanus in the United States in 1777, no immunization existed until the twentieth century. Although German scientists report on tetanus in 1890, a tetanus toxoid, called anatoxin, was not developed until the 1920s. This discovery by French veterinarian Gaston Ramon enabled the immunization of animals against tetanus.

944. Mulder, John B. "Equine Colic: From Curse to Cure." **Veterinary Heritage: Bulletin of the American Veterinary History Society** 14 (September 1991): 84-90.

A brief history of colic, which remains a threat to the health of horses. Changes in treatment over the years are mentioned and references are provided.

945. _____. "Roup Revelations." **Veterinary Heritage: Bulletin of the American Veterinary History Society** 15 (March 1992): 16-22.

A brief history of roup, an upper respiratory tract disease in poultry.

946. Olson, C. and J. Miller. "History and Terminology of Enzootic Bovine Leukosis." In **Enzootic Bovine Leukosis and Bovine Leukemia Virus**, edited by A. Burny, and M. Mammericky, 3-11. Boston: Kluwer Academic Publishers, 1987.

A brief overview of EBL throughout the world since 1871. Includes extensive bibliography.

947. Packer, R. Allen. "News and Notes". **Veterinary Heritage: Bulletin of the American Veterinary History Society**, 15 (March 1992): 23-24.

Provides a list of the veterinary medicine professional organizations that have donated their records to the Archives of American Veterinary Medicine, Department of Special Collections, Parks Library, Iowa State University. The collection emphasizes animal health research, particularly for swine research.

948. Parnas, Joseph. "The Polish Veterinary Science in Dynamic Development: A Historical Remembrance." **Historia Medicinae Veterinariae** [Denmark] 8 (1983): 7-8.

A brief, uncritical note on Polish veterinary medicine. Provides a list of the faculty members who re-organized the veterinary programs in Poland after World War II.

949. Pattison, Iain. **The British Veterinary Profession, 1791-1948**. London: J.A. Allen, 1983. 207 pp.

An excellent history of the veterinary profession including chapters on early veterinary research as well as the Agricultural Research Council. Includes bibliography for materials published after 1828.

950. Pepin, Michel. **Histoire et Petites Histoires des Veterinaires du Quebec**. Montreal, Canada: Editions Francois Lubrina, 1986. 351 pp.

An illustrated history of veterinary medicine in Quebec. Written to celebrate the centennial of the only

French veterinary school in the Americas--Ecole Veterinaire Francaise de Montreal--founded in 1886. Includes sources of manuscript collections.

951. Polydorou, K. "A History of Classical Swine Fever in Cyprus." **British Veterinary Journal** 142 (March/April 1986): 151-54.

A historical survey including a discussion of agricultural policy.

952. Rosenkrantz, Barbara Gutmann. "The Trouble With Bovine Tuberculosis." **Bulletin of the History of Medicine** 59 no. 2 (1985): 155-75.

During the late nineteenth century, scientists disagreed about the relationship between bovine and human tuberculosis. The reform movement to ensure pure meat and milk contributed to that controversy because the veterinary and medical professions each attempted to capture that regulatory area.

953. Rudling, T.A. "Disease Eradication in Tasmania." **Historia Medicinae Veternariae** [Denmark] 8 (1983): 59-61.

The island location and an active veterinary service have enabled the control many animal diseases. All cattle entering Tasmania must go into the state operated quarantine station prior to being released.

954. Ruiz-Martinez, Carlos. **Veterinaria Venezolana: Treinta Anos de Fomento Ganadero, Sanidad Animal e Higiene**

Veterinaria en Venezuela, 1936-1966. Caracas: Editorial Sucre, 1966. 539 pp.

A basic history of veterinary medicine in Venezuela.

955. Saunders, Leon Z. "From Osler to Olafson: The Evolution of Veterinary Pathology in North America." **Canadian Journal of Veterinary Research** 51 (January 1987): 1-26.

Historical data is presented to support the theory that veterinary pathology was first taught in Canada by William Osler (1849-1919), introduced to the United States by Osler's student, Walter Williams (1856-1945) and promoted by Williams' student, Peter Olafson (1897-1985).

956. Schwabe, Calvin W. **Cattle, Priests, and Progress in Medicine**. Minneapolis: University of Minnesota Press, 1978. 277 pp.

Discusses the importance of veterinary studies and a comparative approach to medicine for the promotion of human health. He encourages the use of veterinary programs to help train medical researchers.

957. Shishkov, V.P. "Veterinary Science at the 60th Anniversary of the Soviet Union." **Veterinariia** 12 (1982): 33-36.

An uncritical, brief note on Soviet veterinary medicine.

958. Skrzypek, W. "Prakyczne Wykorzystanie Osiagniec Nauk Weterynaryjnyck w XVIII i XIX Wieku." **Medycyna Weterynaryjna** 39 no. 11 (1983): 695-699.

    Discusses the practical application of the discoveries in veterinary medicine in Poland during the eighteenth and nineteenth centuries.

959. Smith, Frederick. **The Early History of Veterinary Literature and Its British Development**. 4 Vols. London: J.A. Allen, 1976.

    A reprint of the 1919-1933 edition, which covers the history of veterinary medicine until 1860. Emphasizes the British Isles, and it includes name indexes useful for brief, biographical sketches.

960. Smithcors, J.F. **The American Veterinary Profession, Its Background and Its Development**. Ames: Iowa State University Press, 1963. 704 pp.

    The classic work on the history of American veterinary medicine. Smithcors examines the background of the veterinary profession by reviewing the care of livestock from the colonial period until the Civil War. Then, he discusses the profession from 1863-1963. This work was abridged, updated, and published in 1975 as **The Veterinarian in America, 1625-1975**.

961. Stalheim, Ole H.V. "Contributions of the Bureau of Animal Industry to the Veterinary Profession." **Journal of**

American Veterinarian Medical Association 184 (May 15, 1984): 1222-24.

A useful, brief sketch of research contributions by the Bureau. Includes a chronology of disease eradication programs.

962. _____. "The Hog Cholera Battle and Veterinary Professionalism." **Agricultural History** 62 (Spring 1988): 116-21.

With the creation of the Bureau of Animal Industry in 1884, scientists in the Department of Agriculture gave increased attention to controlling animal diseases, one of the most serious of which was hog cholera. In 1905, Marion Dorset, W.B. Niels, and C.B. McBride discovered a serum for hog cholera. Soon vaccination against this disease began to decrease losses for swine producers. By 1978, serum improvements and the use of antibiotics had made the United States free from hog cholera.

963. _____. **Veterinary Medicine in the West**. Manhattan, Kansas: Sunflower University Press, 1988. 82 pp.

A collection of nine essays dealing with developments in veterinary medicine in Canada and the United States during the late nineteenth and early twentieth centuries.

964. Steele, James H. "Veterinary Public Health in the United States, 1776 to 1976." **Journal of the American Veterinary**

Medicine Association 169 (July 1, 1976): 74-82.

A short history of the importance of veterinary medicine to public health. The link between the diseases of man and animals have long been recognized. Provides information about the practitioners and agencies concerned with the relationship between animals and public health.

965. Stowe, Clarence M. "History of Veterinary Pharmacotherapeutics in the United States." **Journal of the American Veterinary Medicine Association** 169 (July 1, 1976): 83-89.

Contends the history of veterinary pharmacology began with treatments used by Native Americans.

966. Tang, Ho Yin. "The Enigma of Hog Cholera: Controversies, Cause, and Control, 1833-1917." Ph.D. diss., University of Minnesota, 1986. 192 pp.

A history of the research conducted by scientists in the United States Bureau of Animal Industry from 1885 to 1917 that led to the development of a swine vaccine to prevent hog cholera.

967. Thierfelder, H. "Tierheilkunde an der Universitat Rintein." **Deutsche Tierarztliche Wochenschrift** 87 no. 4 (1980): 134-137.

A brief review of veterinary science at the University of Rintein, Germany, from 1730 to 1780.

968. Truszczynski, Marion. "Osiagniecia Nauk Weterynaryjnych w 30-leciu Polski Ludowej." **Medycyna Weterynaryjna** 30 no. 7 (1974): 385-389.

An overview of the achievements of veterinary science in Poland since 1945.

969. Weaver, B.M.Q. "The History of Veterinary Anaesthesia." **Veterinary History: Bulletin of the Veterinary History Society** [Great Britain] n.s. 5 no. 2 (Winter 1987/88): 43-57.

A history of the development of veterinary anesthesia with a chronology. The first documented use of anesthesia for British veterinary surgery occurred in London on January 29, 1847, when ether was administered to a horse.

970. West, Geoffrey P., ed. **Black's Veterinary Dictionary**. Totowa, NJ: Barnes & Noble, 1988. 703 pp.

A standard dictionary since 1928. Provides definitions and the history of diseases, including first recorded cases and times of eradication.

971. Whitaker, Adelynne H. "Pesticide Use in Early Twentieth Century Animal Disease Control." **Agricultural History** 54 (January 1980): 71-81.

Discusses the work of the Bureau of Animal Industry's Biochemic and Zoological divisions in the USDA, including the testing of commercial pesticides. Describes the regulatory activity of the USDA until the passage of the Insecticide and Fungicide Act of 1910.

972. Wise, G.H. **Hog Cholera and Its Eradication: A Review of the U.S. Experience**. Animal and Plant Health Inspection Service (APHIS) series no. 91-55. Washington: USDA/APHIS, 1981. 65 pp.

A non-technical overview. Commemorative in nature to those who worked to eradicate the disease. Emphasizes the research that formally began in 1961 until hog cholera was eliminated by 1972.

973. Wiser, Vivian D., Larry Mark, and H. Graham Purchase, ed. **One Hundred Years of Animal Health**. Beltsville, MD: Associates of the National Agricultural Library, 1987. 230 pp.

Surveys the origin and development of the Bureau of Animal Industry and successor agencies in the U.S. Department of Agriculture. Emphasizes the work of the USDA scientists to control and eradicate livestock and poultry diseases and the development of parasitology programs. This was a special issue of the **Journal of NAL Associates** (1986) that includes articles by John S. Andrews, T. C. Byerly, Ole Stalheim, Kay Wheeler and Vivian D. Wiser.

974. Witter, J.F. "The History of Avian Medicine in the United States: I. Before the Big Change." **Avian Diseases** 20 (October-December 1976): 621-30.

A general discussion of the history and classification of avian diseases. Includes Leonard Pearson's work on tuberculosis as well as other research on cholera and pneumonia.

CHAPTER XIV

TECHNOLOGY

GENERAL

975. **Agricultural Engineering**. Rome: Food and Agriculture Organization of the United Nations, 1971. 195 pp.

     An annotated bibliography of FAO publications and documents from 1945 to 1971.

976. Abdel-Daymen, S. "Development of Subsurface Drainage Technology in Egypt." **Proceedings of Symposium on Land Drainage for Salinity Control in Arid and Semi-arid Regions, 1990**, 81-92. Cairo: Drainage Research Institute, 1990.

     Reviews the institutional history of drainage in Egypt. Notes technology, techniques and research.

977. Abraham, M.F. "Appropriate Technology: Basic Needs and Development: Where are the Social Scientists?" In **Sociology of**

Agriculture: Technology, Labour, Development and Social Classes in an International Perspective, edited by A. Bonanno, 77-102. New Delhi: Concept Publishing Co., 1989.

Argues that India has a well-developed research base for technology but little network for diffusion into the countryside. Contends that India's social scientists must become more involved with the spread of technology by encouraging the adoption of appropriate implements.

978. Achour, M.A.B. "The Acceptance and Rejection of Agricultural Innovations by Small Farm Operators: A Case Study of a Tunisian Rural Community." In **Labour, Employment and Agricultural Development in West Asia and North Africa**, edited by D. Tully, 165-89. Dordrecht, Netherlands: Kluwer Academic Publishers, 1990.

A study of the farmers in the northwestern Tunisian village of Lorbous. Contends that technological adaptation depends on income and farm size rather than shortages of farm labor.

979. Amer, M.H., and N.A. de Ridder, eds. **Land Drainage in Egypt**. Cairo: Drainage Research Institute, 1990. 377 pp.

Reviews contemporary land drainage in Egypt as well as historical developments relating to policy, technology, irrigation and agricultural lands.

980. Bainer, Roy. "Science and Technology in Western Agriculture." **Agricultural History** 49 (January 1975): 56-72.

Traces the development of combine and sugar beet harvesters as well as cotton pickers in California. This potpourri article also gives brief attention to tractors, and nut, tomato and grape harvesters and irrigation systems. The footnotes compensate, in part, for the lack of depth in this introductory article.

981. Bayri, T.Y., and W.H. Furtan. "An Economic Analysis of Technological Change in the Spring Wheat Region of Turkey." **ODTU Gelisme Dergisi** 14 no. 4 (1988): 291-313.

Discusses the efforts to introduce high-yield varieties during the late 1960s. Contends that technological change is difficult because it would contribute to unemployment problems. Moreover, improved extension efforts are needed to convince Turkish farmers to adopt new technology.

982. Bell, Jonathan, and Mervyn Watson. **Irish Farming: Implements and Techniques, 1750-1900**. Edinburgh, Scotland: John Donald Pub., 1986. 256 pp.

Analyzes the effects of economic, social and ecological conditions on technological development of Irish agriculture. Argues that retention of traditional cultivation techniques meant survival rather than backwardness.

983. Bidwell, Percy, and John I. Falconer. **History of Agriculture in the Northern United States, 1620-1860**. Washington, D.C.: Carnegie Institution of Washington, 1925. 512 pp.

A basic introduction to technological change in the Northern United States. A dated but useful reference.

984. Barton, Glen T. "Technological Change, Food Needs, and Aggregate Resource Adjustment [1940-58]." Journal of Farm Economics 40 (December 1958): 1429-37.

Discusses the role of technology in regard to surplus production and agricultural adjustment.

985. Busse, W. "Quaderballenpressen: Entwicklung und Stand der Technik." **Landtechnik** 46 no. 4 (1991): 146-49.

Discusses the history and current development of square balers.

986. Chambers, J.D., and G.E. Mingay. **The Agricultural Revolution, 1750-1880**. New York: Schocken Books, 1966. 222 pp.

Emphasizes the major changes in British agricultural technology. A good introduction.

987. **Comprehensive Index of ASAE Publications**. St. Joseph, MI: American Society of Agricultural Engineers, 1971-1984

This annual index continues after 1984 as the **Combined Index to Agricultural Engineering, Transactions of the ASAE,**

ASAE Conference Proceedings. The 1978 volume is titled **Comprehensive Keyword Index of ASAE Publications**.

988. Copestake, J.G. "The Scope for Collaboration Between Government and Private Voluntary Organizations in Agricultural Technology Development: The Case of Zambia." **Network Paper: Agricultural Administration Research and Extension Network, No. 20**, 1990. 41 pp.

Focuses on the Gwembe Valley Agricultural Mission. Contends that technological development is a state responsibility, but that private voluntary organizations can play an active role. Urges more coordination between government and volunteer agencies to improve technological change in Zambia.

989. Czako, Sandor. "A New Tool is Born." **New Hungarian Quarterly** [Hungary] 21 (1980): 148-152.

A study of the rekk, a special hay wagon, developed in northern Moldavia about 1918. Immigrants to the United States brought the design with them.

990. Danhof, Clarence H. **Change in Agriculture: The Northern United States, 1820-1870**. Cambridge: Harvard University Press, 1969. 322 pp.

An excellent study of technological change in relation to plowing, planting and harvesting.

991. Faigmane, L.O. "A Review of Policies and Strategies in Agricultural Mechanization in the Philippines." **RNAM Newsletter** no. 38 (1990): 6-7.

Sketches the historical strategies for mechanization. Notes current agricultural policy in relation to agricultural science and technology.

992. Figueiredo, V. "Small Farmers and Food Production in South Brazil: A Way Out of Imposed Technology." In **Sociology of Agriculture: Technology, Labour, Development and Social Classes in an International Perspective**, edited by A. Bonanno, 169-77. New Delhi: Concept Publishing Co., 1989.

Argues that agricultural development has been shaped by authoritarian governments whose polices reflect corporate needs. Contends that technological change has been imposed on Brazilian farmers.

993. Francks, Penelope. **Technology and Agricultural Development in Pre-War Japan**. New Haven: Yale University Press, 1985. 322 pp.

A extremely useful study of technological change in Japanese agriculture between 1860 and 1940, with the rice industry examined in detail. Good emphasis on the manner in which local communities overcame specific problems that hindered profits.

994. Fussell, G. E. **The Farmers' Tools, 1500-1900: The History of British Farm Implements, Tools and Machinery Before**

the Tractor Came. London: A. Melrose, 1952. 246 pp.

Traces the origin, development and use of agricultural technology in Europe. Emphasis given to British farm tools. An excellent introduction to European agricultural technology.

995. Gates, Paul W. **The Farmers' Age, Agriculture, 1815-1860**. New York: Holt, Rinehart and Winston, 1960. 460 pp.

Although this study is economic in emphasis, it provides a useful introduction to the technology employed in American agriculture during the first half of the nineteenth century.

996. Gray, Lewis Cecil. **History of Agriculture in the Southern United States to 1860**. 2 Vols. Glouster, MA: Peter Smith, 1958.

An excellent introduction to technological change in the American South prior to the Civil War.

997. Grigg, D. **English Agriculture: An Historical Perspective**. Sheffield, United Kingdom: University of Sheffield, 1989. 256 pp.

Examines the causes and consequences of great change in the agricultural practices in England and Wales since World War II. Describes mechanization and plant and animal breeding and relates these changes to the past 300 years of English agriculture.

998. Heikkoinen, Esko. "The Coming of Foreign Agricultural Technology to Finland from the 1870's to World War I." University of Turku, Institute of General History, **Publications** [Finland] 10 (1983): 111-114.

Analyzes the American influence on the mechanization of Finnish agriculture at the turn of the twentieth century, especially for tilling and harvesting. Also includes a discussion of the importance of Swedish sowing and threshing machinery as well as British and Finnish steam engines.

999. Hine, Howard Jordan. **Dictionary of Agricultural Engineering**. Cambridge: W. Heffer, 1961. 252 pp.

A dictionary designed for engineers, farmers and students. Historians of agricultural technology will also find it useful.

1000. Horwith, Bruce J., et al. "The Role of Technology in Enhancing Low-Resource Agriculture in Africa." **Agriculture and Human Values** 6 (Summer 1989): 68-84.

Urges adapting technology to traditional or indigenous agricultural systems. New technology should be designed to help provide food security, ensure conservation, mesh with social constraints and meet financial limitations. More sociological than technological, but it includes a useful bibliography.

1001. Hurt, R. Douglas. **American Farm Tools: From Hand Power to Steam Power**. Manhattan, KS: Sunflower University Press, 1981. 121 pp.

Traces the major features of technological change from the colonial period to the early twentieth century. Topics of study include the origin and development of plows, harrows, grain drills, cultivators, reapers, binders, combines, corn pickers, threshing machines, mowers and fodder-making equipment and steam engines.

1002. _____. **Agricultural Technology in the Twentieth Century**. Manhattan, KS: Sunflower University Press, 1991. 106 pp.

Surveys the major technological developments in the American West during the twentieth century. Subjects include tractors, cotton pickers and strippers, combine harvesters, irrigation equipment, sugar beet harvesters and tomato pickers.

1003. **Information Sources on the Agricultural Implements and Machinery Industry**. New York: United Nations, 1982. 107 pp.

A guide designed for developing countries. Includes a bibliography. Historians of agricultural technology may find this source useful. The introduction and notes are written in English, French, Russian and Spanish.

1004. Jenkins, J.G. **The English Farm Wagon: Origins and Structure**. Newton Abbott, United Kingdom: David & Charles, 1981. 248 pp.

Originally published in 1961, this study describes the construction of 600 wagons used in England, Wales and Ireland.

1005. Kahk, Juhan. "The Spread of Agricultural Machines in Estonia from 1860-1880." **Agricultural History** 63 (Summer 1988): 33-44.

In Estonia, agricultural development lagged, in part, because of imperfect implements, inadequate agricultural science and a traditional distrust of innovation. Social and economic changes during the late nineteenth century, however, stimulated technological change in Estonian agriculture.

1006. Khan, M. "The Process of Technological Change in the Agriculture of a Bangladesh Village: Its Relevance to Mode of Production." **Journal of Social Studies Dhaka** [Bangladesh] no. 53 (1991): 38-67.

Analyzes the influence of technology on labor, credit and marketing at the village level.

1007. Kranzberg, Melvin, and Carroll W. Pursell, Jr., eds. **Technology in Western Civilization**. 2 Vols. New York: Oxford University Press, 1967.

Provides introductory information on the major developments in American agriculture during the twentieth century, including corn, wheat, soybean and livestock breeding and the development of synthetic fertilizer. Technological developments are also included, such as steam traction engines, gasoline tractors, and cotton pickers.

1008. Langdon, John. **Horses, Oxen, and Technological Innovation: The Use of Draught Animals in English Farming from 1066-1500**. New York: Cambridge University Press, 1986. 331 pp.

A history of agricultural and technological change in medieval England, especially innovations in vehicular transport and tillage equipment. Argues the peasantry was the most technologically active class in British agriculture.

1009. Lee, Carol Anne. "Wired Help for the Farm: Individual Electric Generating Sets for Farms, 1880-1930." Ph.D. diss., Pennsylvania State University, 1989. 275 pp.

Farmers began to electrify their farms at the turn of the twentieth century by using portable generators and by linking their farms to commercial power suppliers. The Great Depression and the battle over the control of regional generating plants temporarily halted the electrification of American farms by these methods.

1010. Lehrmann, Joachim. "Probleme Versorgung der Deutchen Landwirtschaft mit Landmaschinen un Geraten im Zweiten Weltkrieg." **Jahrbuch fur Wirtschaftsgeschichte** [German Democratic Republic] 1 (1981): 55-78.

Discusses the supply of Germany's agricultural machinery during World War II. Hitler gave the production of agricultural equipment a high priority before 1939, but wartime shortages of iron and steel ended production.

1011. Madhi, K.A., and D. Tully. "Agricultural Labor and Technological Change in Iraq." In **Labor and Rainfed Agriculture in West Asia and North Africa**, edited by D. Tully, 209-27. Dordrecht, Netherlands: Kluwer Academic Publishers, 1990.

Evaluates the mechanization of dry land farming and its affects on agricultural labor. Primarily social and economic in focus rather than technological.

1012. Maresch, Gerhard. "Die Anfange der Mechanisisierung der Landwirtschaft in Osterreich: Dargestellt Anhand der Sammulung von Modellen Landwirtschaftlicher Gerate und Maschinen des Technischen Museum Wein." **Glatter fur Technikgeschichte** [Austria] (1984/1985): 46-47, 39-80.

An illustrated history of of agricultural mechanization in Austria. Emphasizes the work of Peter Jordan (1751-1827), who is known as the father of Austrian agriculture, and Anton Burg (1767-1849).

1013. Matthews, J. "Farm Power: From Muscle to Microchip." **Journal of the Royal Agricultural Society of England** 149 (1988): 58-70.

Surveys the changes in power sources. Includes human and animal power, tractors and automatic guidance systems. Also discusses harvesting technology, sprayers and fertilizer equipment.

1014. Mingay, G.E., ed **The Agricultural Revolution: Changes in Agriculture, 1650-1880**. London: Adams & Charles Black, 1977. 322 pp.

A collection of essays that stress the economic aspects of technological change in English agriculture. Includes chapters on livestock, machinery and draining as well as economic matters, such as capital and labor.

1015. Morgan, Raine. **Farm Tools, Implements and Machines in Britain: Pre-History to 1945**. Reading, England: Institute of Agricultural History, University of Reading, 1984. 275 pp.

A useful bibliography, but it is not annotated. Includes subject and author indexes. Although the emphasis is on Britain, general works that include other countries are included. The section on manufacturer's catalogues is particularly useful.

1016. Niederer, Arnold. "Die Alpine Alltagskultur: Zwischen Routine und der Adoption con Neuerungen." **Schweizerische Zeitschrift fur**

Geschichte [Switzerland] 29 no.1 (1979): 233-255.

Review of the adoption of agricultural technology in isolated alpine agricultural communities since the fourteenth century, especially for grain production and cheese making. Because these communities are isolated and self-sufficient, decisions about the adoption of new mechanical technology are usually based on whether it will make a job easier rather than on the potential for increased production.

1017. O'Brien, P.K. "Agriculture and The Industrial Revolution." **Economic History Review** 30 no. 2 (1977): 166-181.

A critical analysis of the most important works on British agricultural history. Economic emphasis, but some analysis of agricultural technology and farming techniques.

1018. Orczyk, Jozef. "Problemy Techniki Rolniczej w Polsce w Latch Wielkiego Kryzysu, 1929-1935." **Roczniki Dziejow Spolecznych i Gospodarczych** [Poland] 30 (1969): 123-166.

Drops in farm prices during the Great Depression caused a shortage of capital for the adoption of new agricultural technology. This factor caused greater technological disparity between Poland and the more industrialized countries.

1019. Panayiotou, G., and D. Tully. "Agricultural Labor and Technological Change in Cyprus." In **Labor and Rainfed**

Agriculture in West Africa and North Africa, edited by D. Tully, 135-61. Dordrecht, Netherlands: Kluwer Academic Publishers, 1990.

A detailed description of mechanization, irrigation, improved crops and livestock and research programs. Evaluates the effects of technological change on labor and cropping patterns.

1020. Przedpelski, Jan. "Pionierzy Mechanizacji Rolnictwa na Mazowszu." **Biuletyn Zdowskiego Instytutu Historycznego w Polsce** [Poland] 85 (1973): 79-83.

A history of a Polish factory founded by Mojzesj Sarha that has produced agricultural implements for more than a century. It survived German occupation and continued to operate as a state enterprise.

1021. Pineiro, Martin, and Eduardo Trigo, ed. **Technical Change and Social Conflict in Agriculture: Latin American Perspectives**. Boulder: Westview Press, 1983. 248 pp.

Six case studies of technological change. Theoretical rather than historical, but it merits the attention of historians.

1022. Richards, Alan. "Agricultural Technology and Rural Social Classes in Egypt, 1920-1939." **Middle Eastern Studies** [Great Britain] 16 no.2 (1980): 56-83.

A continuation of Richards' earlier study of Egyptian agricultural technology and its effect on farmers. This study centers on the activities between the two world wars.

1023. _____. "Technical and Social Change in Egyptian Agriculture: 1890-1914." **Economic Development and Cultural Change** 26 no. 4 (1978): 725-745.

The British advocated intensive agriculture in Egypt at the turn of the twentieth century by promoting perennial irrigation and crop rotation. Examines the social and ecological consequences of British agricultural policy in Egypt.

1024. Rogin, Leo. **The Introduction of Farm Machinery in Its Relation to the Productivity of Labor in the Agriculture of the United States During the Nineteenth Century**. Berkeley: University of California Press, 1931. 260 pp.

An important study of plows, harrows, reapers, threshing machines and seeders. Relates these technological developments to productivity and man-hours in American agriculture.

1025. Roy, Sumit. "Agrarian Crisis and Technology in Nigeria." **Africa Quarterly** [India] 25 nos. 1/2 (1986): 1-12.

During the mid-1960s, Nigerian agricultural production decreased because of drought and government policy. Nigeria did not stress the importance of innovative agricultural technology until 1981.

1026. Rozenbaum, Aleksandr Natanovich. **Anglo-Russkii Slovar' po Sel'skokhoziaistvennoi Tekhnike**. Moskva: Sovetskaia Entsiklopediia, 1965. 379. pp.

A useful dictionary for agricultural machinery, published in both Russian and English.

1027. Sangwan, Satpal. "Indian Response to European Science and Technology, 1757-1857." **British Journal for the History of Science** [Great Britain] 21 no. 2 (1988): 211-232.

When the British began their rule of India in 1757, they began to introduce agricultural technology. Farmers were suspicious of Western technology because not all innovations were appropriate for India's soils or crops.

1028. Seyf, Ahmed. "Technical Changes in Iranian Agriculture, 1800-1906." **Middle Eastern Studies** [Great Britain] 20 no. 4 (1984): 142-54.

Based on trade reports and other government publications. The author suggests that Iranian agriculture became less productive throughout the nineteenth century because fertilizer declined in quality and supply.

1029. Simpson, M.J. "Researching Old Farm Machinery in Australia." Conference on Agricultural Engineering, Proceedings of a Conference Held at Hawkesbury, Richmond, NSW, Australia, 25-30

September 1988, **National Conference Publication No. 88/12**, (1988): 5-8.

Describes visual and published evidence for the identification of Australian farm equipment used from 1880 to 1930.

1030. Singer, Charles, et al. **A History of Technology**. 8 Vols. Oxford: Clarendon Press, 1954-1984.

A thorough introduction to European agricultural technology from antiquity to approximately 1950. Volume 8 is a comprehensive index. Anyone coming to this field for the first time should begin here.

1031. Steinmetz, Heinrich. **Landmaschinen und Gerate: Mehrsprachen-Bildworterbuch**. Betzdorf: H. Steinmetz, 1976. 476 pp.

A multi-lingual dictionary for agricultural implements.

1032. Stoyanov, K.H. "Developments and Achievements of Agricultural Science in the Field of Mechanization of Agriculture in the Last 25 Years." **Selskostopanska Tekhnika** [Bulgaria] 26 (1989): 1, 5-27.

A survey of technological improvements in Bulgarian agriculture since the mid-1960s, such as tractors, fertilizers, planters and cultivators.

1033. **The Adoption of Technology in Agriculture**. 6 Vols. London: National Economic Development Office, 1986.

A detailed survey of agricultural innovation and technological change in Great Britain. Topics include silage, poultry, cereals, oilseed rape, glasshouses and sheep.

1034. Thomas, T.H., and A.M. Hawkins. "Animal-derived Shaft Power for Rural Africa." **Energy and the Environment: Into the 90s, Proceedings of the 1st World Renewable Energy Congress, Reading, U.K., 23-29 September 1990**, edited by A.A.M. Sayigh, 2070-2074. Oxford: Pergamon, 1990.

Describes historical and contemporary gear designs for animal-powered machines intended to improve the stationary work of draft animals for lifting water, milling grain and generating electricity.

MECHANIZATION

1035. Agarwal, Bina. **Mechanization in Indian Agriculture: An Analytical Study Based on the Punjab**. New Delhi: Delhi School of Economics, 1983. 290 pp.

A study of 240 farms in the wheat growing region of the Punjab during the early 1970s. Emphasizes the effect of tractors on crop production and farm employment. Statistical methodology.

1036. Ankli, Robert E., H. Dan Helsberg, and John H. Thompson. "The Adoption of the Tractor in Western Canada." **Canadian Studies in Rural History** 2 (1979): 9-39.

Analyzes the revolutionary effects of the tractor on Canadian agriculture from 1920 to 1950. Evaluates the acceptance of this new technology by farmers.

1037. Binswanger, H.P. "Agricultural Mechanization: A Comparative Historical Perspective." **World Bank Staff Working Paper, International Bank for Reconstruction and Development No. 673**, 1984. 80 pp.

Compares technological adoption in developed and undeveloped countries. Contends that mechanization is most successful where farms are large and human labor is insufficient. Mechanization came first to power intensive operations, such as pumping and processing, and later to intensive operations when the economic returns merited it.

1038. Chamberlain, A.H. "The Chamberlain Tractor Story." **Agricultural Engineering Australia** 17 no. 1 (1988): 2-11.

An overview of agricultural engineering in Australia in relation to tractor development.

1039. Cloud, Gayla Staples. **Agricultural Mechanization in the Third World: A Selected Bibliography, 1975-1985**. Monticello, IL: Vance Bibliographies, Public Administration Series p-2029, 1986. 41 pp.

A good introduction to English publications or those with English summaries. Analyzed by continent.

Excludes the literature on specific crops unless it is primary for a certain area.

1040. Collins, E.J.T. "The Rationality of 'Surplus' Agricultural Labor: Mechanization in English Agriculture in the Nineteenth Century." **Agricultural History Review** [Great Britain] 35 pt. 1 (1987): 36-46.

Discusses agricultural unemployment, labor productivity and migration to urban areas in relation to the mechanization of agriculture in the United Kingdom.

1041. Culpin, Claude. **Farm Machinery**. St. Albans: Granada, 1981. 450 pp.

The classic descriptive study of British agricultural implements. Written for farmers and students, but useful for historians. Includes irrigation and crop drying equipment as well as implements for dairying, silage making and livestock raising. The older editions are particularly helpful for the study of machines that are no longer manufactured but which are still used in developing countries.

1042. Desert, Gabriel. "Machinisme et Agriculture dans la France de XIXe Siecle." **Historical Papers** [Canada] (1984): 185-216.

During the nineteenth century, France failed to adopt new agricultural technology and lagged in productivity compared to neighboring countries.

French farmers were too conservative and implement manufacturers too lacking in entrepreneurial abilities.

1043. Espeli, H. "Fra Hest Til Hestekrefter: Studier i Politiske og Okonomiske Rammebetingelse for Mekaniseringen av Norsk Jordbruk, 1910-1960." **Melding: Institutt for Okonomi og Samfunnsfag, Norges Landbrugshogskole**. As, Norway: Institutt for Okonomi og Samfunnsfag, Norges Landbrugshogskole, 1990. 849 pp.

Discusses agricultural mechanization in Norway from 1910 to 1960 in relation to public policy and business. Emphasizes policy rather hardware.

1044. Gray, R.B. **The Agricultural Tractor, 1855-1950**. St. Joseph, MI: American Society of Agricultural Engineers, 1975. 63 pp.

An excellent chronological introduction to the technical development of the tractor in the United States. Well illustrated.

1045. Herrer, J. de. "Agricultura y Technologia: Historia Moderna de la Mecanizacion Agraria." **Maquinas y Tractores Agricolas** 1 no. 1 (1990): 74-79.

Surveys Spanish agricultural mechanization in relation to the European Economic Community. Covers the period 1960-1988, with emphasis on the use of tractors.

1046. **History, Facilities, Organization and Research Triennial Report (1986-88)**. Bologna: Institute of Farm Machinery and Agricultural Mechanization, University of Bologna, 1989. 70 pp.

Historians of agricultural technology will find this study useful for contemporary information about mechanical harvesting, pesticides, tillage and environmental pollution.

1047. Holt, Pliny E. "The Development of the Track-Type Tractor." **Agricultural Engineering** 6 (April 1925): 76-79.

A brief but useful review of the early work to develop this form of tractor.

1048. L'vunin, Iu A. "Traktor-Mezhrabpom na Urale, 1923-1928." **Istoriia SSR** [Soviet Union] no. 1 (1981): 137-144.

A study of the International Workers' Relief joint-stock tractor company founded in 1923 with the aid of a small group of American Communists. This company became a major producer of tractors in the Southern Urals.

1049. Larsen, L. **Farm Tractors, 1950-1975**. St. Joseph, MI: American Society of Agricultural Engineers, 1981. 184 pp.

Updates the history of tractors by R.B. Gray. Annual chronology of major developments.

1050. Nekrasov, S. and I. Boichuk. "Vosproizodstvo Sroki Sluzhby Sel'Skokhoziaistvennoi Tekhniki."

Voprosy Ekonomiki [Soviet Union] 8 (1986): 73-81.

A critical analysis of the durability of Soviet tractors and combine harvesters since the mid-1950s.

1051. Nemeth, David J. "The Walking Tractor: Trojan Horse in the Cheju Island Landscape." **Korean Studies** 12 (1988): 14-38.

This study examines the effects of the walking tractor on the South Korean village-centered, agrarian system during the 1970s and 1980s. Traditionally, farmers on Cheju Island used beasts of burden, but this new technology began to eliminate the need for these animals.

1052. Norbeck, J. **Encyclopedia of American Steam Traction Engines**. Sarasota, FL: Crestline Publishing Co., 1976. 320 pp.

Written for the steam engine buff, this study contains valuable technical information that the scholar will find useful.

1053. Poel, J.M.G. van der. "A Hundred Years Agricultural Mechanization in the Netherlands." **Acta Historiae Neerlandica** [Netherlands] no. 5 (1971): 316-325.

By the mid-nineteenth century, the Dutch began to import plows from the United States. Information on U.S. agricultural technology was transferred between relatives in America and the Netherlands. This article summarizes other works by Poel on the mechanization of Dutch agriculture, including dairying.

1054. Povilyunas, A.F. "Istoriya Razvitiya Sel'skogokhozyaistva Litovskoi SSR--Istoriya Sistem Ego Vedeniya." **Vestnik Sel'skokhozyaistvennoi Nauki Kazakhstana** 1989 no. 9 (1989): 16-20.

Discusses agricultural management in Lithuania during the past thirty years. Not specifically related to mechanical hardware but useful for understanding the problems of applying mechanization to Lithuanian agriculture.

1055. Quick G.R. "The Australian Tractor Industry: Perspectives and Lessons from an Overmature Market." In **Supplement to the Proceedings of the International Symposium on Agricultural Engineering, Potentialities of Agricultural Engineering in Rural Development, Beijing, China, 12-15 September 1989**, edited by M.H. Wang, 11-14. Beijing: Chinese Society of Agricultural Engineering, 1989.

Tractor production began in 1909 with twenty-five manufacturers. By the late 1980s, only six remained and none engaged in mass production. Shows the linkage of agricultural profits to mechanization.

1056. Rackham, D.H., and D.P. Blight. "Four-wheel Drive Tractors: A Review." **Journal of Agricultural Engineering Research** 31 no. 3 (1985): 185-201.

Discusses the history of 4-wheel drive tractors to the 1960s. Notes the differences in steering design and performance characteristics.

1057. Rijk, A.G. **Agricultural Mechanization Policy and Strategy: The Case of Thailand**. Tokyo: Asian Productivity Organization, 1989. 283 pp.

Examines the history of agricultural mechanization in Thailand, but much of this study is based on theoretical model building.

1058. Scheinman, D. **The Feasibility of Tractor Mechanization in Tanga Region**. Tanga, Tanzania: Tanga Integrated Rural Development Program, 1989. 206 pp.

Discusses tractor usage in Tanzania and notes the history of agricultural mechanization. Economic in focus.

1059. Shepard, R. Bruce. "Tractors and Combines in the Second State of Agricultural Mechanization on the Great Plains." **Prairie Forum** [Canada] 11 (Fall 1986): 253-71.

Discusses the evolution of agricultural technology on the Canadian Great Plains from the late nineteenth to the mid-twentieth centuries, especially for the acceptance and adoption of the tractor and combine.

1060. Sigaut, F. "L'innovation Mecanique en Agriculture: Essai d'une Analyse Historique Comparative. **Cahiers de la Recherche Development** no 21 (1989): 1-9.

Reviews the mechanical developments in agriculture before the Industrial Revolution. Emphasizes planting and threshing, with special reference to Africa.

1061. **Utilization of Farm Machinery in Asia: Report of APO Multi-Country Study Mission, 19th-29th June, 1990**. Tokyo: OQEH, 1991. 302 pp.

Describes mechanization in the United States, Western Europe and Asia, with special reference to Japan. See also for references to mechanization in Brunei, Taiwan, India, Indonesia, the Republic of Korea, Malaysia, Nepal, Pakistan, Sri Lanka, Thailand, Tonga and the Philippines.

1062. Vasilenko, P.M., and P.P. Panchenko. **Razvitie Mekanizatsii i Elektrifikatsii Sel'skogo Khoziaistva Ukrainskoi SSR**. Kiev: Dunka, 1988, 470 pp.

A history of the mechanization and electrification of Ukranian agriculture.

1063. Vati, L., and L. Zentai. "A Magyar Mezogazdasag Gepesitese es Munkaerogazdalkodasa." **Gazdasag es Statisztika** [Hungary] 2 no. 3 (1990): 41-56.

Discusses the mechanization of Hungarian agriculture since 1960. Economic and social rather than hardware in focus.

1064. Wendel, C.H. **Encyclopedia of American Farm Tractors**. Sarasota, FL: Crestline Publishing Co., 1979. 352 pp.

Written for the tractor buff, but it contains a great deal of technical information that the scholar will find useful and difficult to locate in other sources.

1065. Wik, Reynold Millard. "Henry Ford's Science and Technology for Rural America." **Technology and Culture** 3 (Summer 1962): 247-58.

An evaluation of Ford's efforts to apply science and technology to American agriculture, including attempts to make fuel alcohol from grain, rubber from weeds and synthetic milk from soybeans. Concludes that Ford was ahead of his time.

1066. _____. **Steam Power on the American Farm**. Philadelphia, 1953. 288 pp.

An excellent technological and economic history of the development of steam engines for American agriculture.

1067. Williams, M. **British Tractors for World Farming: An Illustrated History**. Poole, United Kingdom: Blandford Press, 1980. 112 pp.

A popular history of the tractor industry during the past eighty years.

1068. _____. **Steam Power in Agriculture**. Poole, United Kingdom: Blandford Press, 1977. 183 pp.

A brief history of stationary and self-propelled steam traction engines in Great Britain.

1069. Williams, Robert C. **Fordson, Farmall, and Poppin' Johnny: A History of the Farm Tractor and Its Impact on America**. Urbana: University of Illinois Press, 1987. 232 pp.

The most comprehensive history of the the farm tractor in America. Contains considerable information about the social and economic influence of the tractor as well as hardware developments.

TILLAGE

1070. Amanor, Kojo Sebastian. "Managing the Fallow: Weeding Technology and Environmental Knowledge in the Krobo District of Ghana." **Agriculture and Human Values** 8 (Winter/Spring 1991): 5-13.

The title is misleading. This study is more concerned about indigenous cropping patterns than technological hardware. Urges the study of indigenous agricultural knowledge.

1071. Bartels, R. "Ruckblick auf 110 Jahre Feldversuche im Bodentechnologischen Institut." **Zeitschrift fur Kulturtechnik und Landentwicklung** 31 no. 2 (1990): 67-71.

A history of the experiments in soil management conducted at the Institute of Soil Technology in Bremen, Germany, during the last 110 years.

1072. Friedman, M., and Z. Tempir. "Geometric Analysis of the Shares of Historical Ploughs with the Breaking-up type of Mouldboard and their Working Characteristics." In **Physical Properties of Agricultural Materials and Products. Proceedings of the 3rd International Conference, August 19-23, 1985, Prague, Czechoslovakia**, edited by R. Rezicek, 999-1004. New York: Hemisphere Publishing Corp., 1988.

Analyzes the shape of four plows in the Prague Agricultural Museum to determine their tillage ability. Useful for understanding field preparation, plowing rates and technological development.

1073. Hall, Alfred. **Ploughman's Progress**. Ipswich, United Kingdom: Farming Press, 1992. 147 pp.

A brief discussion of plows, plowing and tillage history. British and European in focus.

1074. Isern, Thomas D. "The Discer: Tillage for the Canadian Plains." **Agricultural History** 62 (September 1989): 79-97.

Brief history of tillage technology in Canada. Discusses disc harrows and one-way disc plows. Notes the work of scientists Evan A. Hardy and Hartford A. Lewis.

1075. McKyes, E. **Soil Cutting and Tillage**. Amsterdam: Elsevier, 1985. 217 pp.

Surveys the history of tillage practices and implements. Includes a chapter on tractors and discusses the affect of tillage implements on plant growth.

1076. Olmstead, Alan L., Bruce F. Johnston, and Brian G. Sims. "Forward to the Past: The Diffusion of Animal Powered Tillage Equipment on Small Farms in Mexico." **Agricultural History** 60 (Winter 1986): 62-78.

Contends the United States should manufacture and provide animal-powered equipment rather than advanced agricultural technology to developing nations. Farmers in those nations do not have the technological skills or the financial means to adopt modern implements. Instead, an intermediary stage of technological adoption is needed. This study covers the period from 1970 to 1986.

1077. Park, H.S., and H.K. Song. "Origin and Evolution of the Oriental and Occidental Plow (2): Evolution of the Chinese and Indian Plow." **Research Reports of the Rural Development Administration, Agricultural Engineering and Farm Products Utilization** 31 no. 1 (1989): 25-35.

Discusses the history of the Chinese triangular and quadrilateral plows of the Han and Tang dynasties and the Indian stick plow that developed in Afghanistan with influences from Mesopotamia.

1078. _____. "Origin and Evolution of the Oriental and Occidental Plow (3): Evolution of the Korean Plow."

Research Reports of the Rural Development Administration, Agricultural Engineering and Farm Products Utilization 31 no. 1 (1989): 36-51.

Discusses the transformation of the Korean plow from vertical to horizontal to short bottom designs. Covers the period from approximately 200 B.C. to the early twentieth century.

1079. _____. "Origin and Evolution of the Oriental and Occidental Plow (4): Evolution of the Occidental Plow." **Research Reports of the Rural Development Administration, Agricultural Engineering and Farm Products Utilization** 31 no. 3 (1989): 28-46.

Traces the development of the plows used in Western Europe to Egypt and Mesopotamia. A moldboard plow was not developed until the late Middle Ages and the modern European plow did not originate until the seventeenth and eighteenth centuries.

1080. Pratley, J.E., and D.L. Rowell. In **From the First Fleet--Evolution of Australian Farming Systems**, edited by P.S. Cornish and J.E. Pratley, 2-23. Melbourne: Inkata Press, 1987.

A history of plows, discs and seed drills in Australia.

1081. Rajaram, G., D.C. Erbach, and D.M. Warren. "The Role of Indigenous Tillage Systems in Sustainable Food Production." **Agriculture and Human Values** 8 (Winter/Spring 1991): 149-155.

Contends that many farmers in Western countries have adopted limited tillage practices, called conservation tillage, from agriculturists in developing nations. Compares tillage practices in the United States and India and concludes that indigenous tillage practices should be used in the West to help prevent soil erosion and enable sustainable agriculture.

1082. Shi, L.P. "A Brief Introduction on the Development of Boat Type Tillage Machine." **Transactions of the Chinese Society of Agricultural Machinery** 21 no. 2 (1990): 1-6.

Discusses the history of tillage machines designed for paddy fields. Notes the current usage.

1083. Shirwa, A.H., et al., "Ox-Cultivation in Somalia: Historical Background and Present Situation." In **Rural Development Aspects in Eastern Africa, DAAD Seminar, Nairobi/Njoro No. 20**, edited by H.U. Thimm, 131-54, 1990.

Reviews the use of draft animals and tillage implements in Somalia. Notes that most farmers depend on hired tractor services. Good historical survey.

1084. Symes, Oliver L. "Evolution of Tillage Practices for Sustained Crop Production in the Palliser Triangle." **Transactions of the Royal Society of Canada** 1 (1986): 97-104.

This study evaluates land use between the north branch of the Saskatchewan River, the United States and the Rocky Mountains from the early nineteenth to the mid-twentieth centuries. Discusses the use of tractors and disc plows, but stresses the dominance of the environment.

HARVESTING

1085. Abouchar, Alan, and Daniel Needles. "Custom Combining: A Neglected Opportunity for Soviet Agriculture." **Canadian Slavonic Papers** 17 no. 1 (1975): 65-75.

Discusses the reasons for the failure of custom combining in the Soviet Union. Institutional emphasis but good for comparative purposes concerning the use of technology in an innovative manner.

1086. Huxley, B. **Combines**. London, United Kingdom: Osprey Publishing, 1990. 128 pp.

A brief history of the combine. International emphasis. Covers the period 1836 to the present. Illustrated and intended for the introductory student.

1087. Isern, Thomas D. **Bull Threshers and Bindlestifs: Harvesting and Threshing on the North American Plains**. Lawrence: University Press of Kansas, 1990. 248 pp.

An excellent history of the technological innovations for harvesting grain in the Great Plains region of Canada and the United States.

1088. Kolomiichenko, I.I. "Sozdanie Sovetskogo Kombainostroeniia." **Istoriia SSSR** [Soviet Union] 7 no. 6 (1963): 104-114.

A history of the Soviet combine manufacturing industry from 1929 to 1932. Prior to this period combines had been imported from the United States, but by 1932 Soviet industry had built 10,000 combines, and it no longer needed to import American made implements.

1089. Korhonen, Teppo. "Zur Geschichte Der Finnischen Sicheln." **Finskt Museum** [Finland] 91 (1984): 80-108.

Traces the history of the Finnish sickle from A.D. 200 to 1900. The bent sickle was used to harvest rye. Variations in sickles are related to their intended use.

1090. Korol'kov, N.V. "Kombain v Kolkhoznoi Derevne 30-KH Godov: Osvoenie, Sotsial'Nye Posledstviia." **Istoriia SSSR** [Soviet Union] no. 2 (1977): 95-104.

A study of combine harvesters on collective farms in the Soviet Union from the 1920's to the 1930's. During the first Five-Year Plan, combines brought mechanization to Soviet agriculture. Production increased rapidly between 1930 and 1938. This new technology created jobs for peasants, but it required a major transformation in the work force regarding the organization and training of agricultural labor.

1091. MacDonald, Stuart. "The Progress of the Early Threshing Machine." **Agricultural History Review** [Great Britain] 23 no. 1 (1975): 63-77.

An analysis of the advantages and disadvantages of the threshing machine in England. Introduced in 1786, by 1830 it had diffused throughout England. Contends the threshing machine was more useful to farmers in northern England where labor shortages prevailed. In the south, a plentiful labor supply made harvesting by traditional methods less expensive.

1092. McLean, I.W. "The Adoption of Harvest Machinery in Victoria in the Late Nineteenth Century." **Australian Economic History Review** 13 no. 1 (1973): 41-56.

Analyzes the factors contributing to the adoption rate of wheat harvesting technology from 1860-1990.

1093. Pomfret, Richard William Thomas. "The Introduction of the Mechanical Reaper in Canada, 1850-70: A Case Study in the Diffusion of Embodied Technical Change." Ph.D. diss., Simon Frasier University, 1974. 101 pp.

A general history of the technological diffusion and the engineering of the mechanical reaper in Canada.

1094. Quick, Graeme R., and George F. Montogmery. **Bibliography on Combines and Grain Harvesting: Citations from the International Literature on the Engineering, Biological, and Economic Aspects of the Harvesting of Crops for Grain and Seed**. St. Joseph, MI: American Society of Agricultural Engineers, 1974. 71 pp.

An essential bibliography. International in emphasis. An excellent place to begin.

1095. Quick, Graeme, and Wesley Buchele. **The Grain Harvesters**. St. Joseph, MI: American Society of Agricultural Engineers, 1978. 278 pp.

A history of grain harvesting from the sickle in the ancient Middle East to modern combines. Technical developments in the United States and Australia are emphasized. A good introductory source.

1096. Rikoon, J. Sanford. **Threshing in the Midwest, 1820-1940: A Study of Traditional Culture and Technological Change**. Bloomington: Indiana University Press, 1988. 214 pp.

Essentially a social history of threshing, but this study includes some technological discussion. Important for understanding the social process in relation to technological change.

1097. Walker, Thomas S., and G.K. Kshirsagar. "The Village Impact of Machine Threshing and Implications for Technology." Development in the Semi-Arid Tropics of

Peninsular India." **Journal of Development Studies** [Great Britain] 21 no. 2 (1985): 215-31.

Reviews the introduction of machine threshing on the local economy from 1975 to 1980. Economic emphasis.

1098. Wendt, Manfred. "'Zentrale Erntetechnik'--Jugendobjekt der FDJ, 1966-1967 and 1973-1979." **Beitrage zur Geschichte der Arbeiterbewegung** [German Democratic Republic] 31 no. 3 (1989): 360-368.

The East Germans established an experimental program in 1966 to get young people interested in agriculture by training participants to harvest grain using the newest technology. Officials again conducted the experiment from 1973-1979 with mixed results.

IRRIGATION

1099. Adams, William M., and David M. Anderson. "Irrigation Before Development: Indigenous and Induced Change in Agricultural Water Management in East Africa." **African Affairs** [Great Britain] 87 no. 349 (1988): 519-535.

Compares indigenous to modern irrigation systems in Kenya and Tanzania during the 1970s and 1980s. Supports the flexibility and expansion of indigenous irrigation systems.

1100. Ambler, J.S. "Adat and Aid: Management of Small-Scale Irrigation in West Sumatra, Indonesia." Ph.D. diss., Cornell University, 1990. 615 pp.

Describes the irrigation tradition of the Minangkabau people in West Sumatra. Although the emphasis is placed on social structure, this study provides an important analysis of changing policy for water distribution that places authority with the state rather than the individual or community.

1101. Dorji, T. "Farmer-Managed Irrigation Systems in Bhutan." In **Design Issues in Farmer-Managed Irrigation Systems**, edited by R. Yoder and J. Thurston, 271-73. Colombo, Sri Lanka: International irrigation Management Institute, 1990.

The history of irrigation in Bhutan began about twenty years ago with the establishment of an irrigation division in that nation's Department of Agriculture. Discusses the difficulties of irrigating land in mountainous terrain and government policy to support that work.

1102. Fiorentinio, Raul. "Apuntes Para Una Estrategia de Desarrollo de la Agricultura de Riego en la Argentina." **Desarrollo Economico** [Argentina] 27 no. 108 (1988): 539-558.

Describes the development of irrigation in Argentina from 1900 to the 1970s. During the 1970s irrigation declined due to lack of markets for crops, insufficient credit, inadequate technical assistance and planning problems.

1103. Francks, P. **Technology and Agricultural Development in Pre-War Japan**. New Haven: Yale University Press, 1984. 322 pp.

Traces the technological and economic changes in irrigated rice farming on the Saga Plain of Japan from 1850 to 1940. Includes a general discussion of pre-World War II Japanese agriculture. Emphasizes irrigation technology.

1104. Green, Donald E. **Land of the Underground Rain: Irrigation on the Texas High Plains, 1910-1970**. Austin: University of Texas Press, 1973. 295 pp.

An excellent discussion of the technological and economic development of irrigation with underground water on the Southern Great Plains of the United States. An essential source.

1105. Groenfeldt, David. "Building on Tradition: Indigenous Irrigation Knowledge and Sustainable Development in Asia." **Agriculture and Human Values** 8 (Winter/Spring 1991): 114-120.

Surveys indigenous irrigation systems and engineering. India, Pakistan, Sri Lanka and the Philippines are used as examples. Acknowledges the importance of modern technology but urges a recognition of the importance of local knowledge about irrigation to promote sustainable agriculture.

1106. Ibatov, M. "Milioratіvnoe Stroitel'Stvo v Turkmenskoi SSR na Sovremennom Etape" **Istoriia SSSR** [Soviet Union] no.6 (1982): 127-131.

History of the irrigation and land reclamation projects started by the Russians in the Soviet portion of Turkestan after 1918. Notes the substantial increase of irrigated lands since 1965.

1107. **Irrigation and Drainage in China**. Beijing: China Water Resources and Electric Power, 1987. 187 pp.

A brief history of irrigation, drainage and water resources in China. Notes current projects and research.

1108. Iskenderov, I., K.Z. Azizov, and A.A. Mamedov. "History of Development of Irrigation in Arid Zones of Azerbaijan." **Problems of Desert Development, No. 2**, 67-70, 1990.

A brief survey that emphasizes soil salinization, groundwater problems and the construction of irrigation and drainage systems.

1109. Latz, Gil. **Agricultural Development in Japan**. Chicago: University of Chicago Press, 1989. 135 pp.

A brief history of land reclamation and irrigation development in Japan.

1110. Le Moigne, G., S. Barghouti, and H. Plusquellec, ed. "Technological and Institutional Innovation in Irrigation."

World Bank Technical Paper, No. 94, 1989. 141 pp.

Includes twelve essays that evaluate current irrigation technology and its use around the world. Notes research issues and priorities. Some discussion of technological transfer by public and private institutions regarding extension.

1111. Nye, Ronald Loren. "Visions of Salt: Salinity and Drainage in the San Joaquin Valley California, 1870-1970." Ph.D. diss., University of California-Santa Barbara, 1986. 350 pp.

A history of drainage and salinity problems in the San Joaquin Valley of California. Emphasizes technological adaptation to solve the salt problems, although much of this study concerns institutional and economic developments.

1112. Pretty, J.N. "Farmers' Extension Practice and Technology Adaptation: Agricultural Revolution in 17-19th Century Britain." **Agriculture and Human Values** 8 (Winter/Spring 1991): 132-48.

Discusses irrigation, drainage and technological transfer as well as soil fertility, livestock raising and crop production in relation to educational activities.

1113. Raynaut, C. "Irrigated Agriculture in Hausa Areas of Niger: Historical, Social and Technical Aspects." **Etudes Rurales** no. 115-116 (1989): 105-28.

Describes the irrigation technology and techniques in the Maradi region.

1114. Reddy, M.S. "Inter-Basin Water Transfers in India for Regional Irrigation Development." In **Proceedings International Commission on Irrigation and Drainage. Special Technical Session, Beijing, China. Vol. 1-A, Irrigation Planning**, 56-77, 1991.

Discusses the storage of water and its transfer over long distances since historical times.

1115. Rosenthal, J.L. "The Development of Irrigation in Provence, 1700-1860: The French Revolution and Economic Growth." **Journal of Economic History** 50 (September 1990): 615-38.

An economic history that will provide a useful perspective for anyone studying the technological development of irrigation in France.

1116. Soenaro, I. "A New Approach to Development Operation and Maintenance of Irrigation Systems in Indonesia." In **Irrigation: Theory and Practice**, edited by J.R. Rydzewski and C.F. Ward, 869-874. Southampton, United Kingdom: Institute of Irrigation Studies, Southampton University, 1989.

Discusses the recent history of irrigation in Indonesia. Notes technical problems.

1117. Stone, Ian. **Canal Irrigation in British India: Perspectives on Technological Change in a Peasant Economy**. New York: Cambridge University Press, 1984. 374 pp.

Emphasizes the western portion of present-day Uttar Pradesh. Discusses the social adjustment to technological change and places the development of the Doab canals within the context of irrigation agriculture in India.

1118. Underhill, H. **Small Scale Irrigation in Africa in the Context of Rural Development**. Bedford, United Kingdom: Cranfield Institute of Technology, 1990. 90 pp.

Analyzes the problems and possibilities for the development of irrigation in Africa. Provides a survey of past activities. Economic in emphasis but a useful overview for studies of technology.

1119. Whitcombe, Elizabeth. "Irrigation and Railways." In **The Cambridge Economic History of India: Vol. 2, c. 1757-c.1970**, edited by Dharma Kumar and Meghnad Desai, 677-761. New York: Cambridge University Press, 1983.

Traces the development of canal irrigation since 1803 with the work of the East India Company in the kingdoms of Delhi and Tanjore. Notes the limits of irrigation. Economic and institutional in focus.

1120. Yukawa, K. "Historical Development of Rice Cultivation and Its Water Management." **Journal of Irrigation Engineering and Rural Planning** no. 16 (1989): 60-70.

Reviews two centuries of rice cultivation and irrigation in Japan.

TECHNOLOGICAL TRANSFER

1121. Abzari, Mehdi. "Technology Transfer to Less Developed Countries: A Study of Agricultural Development in India Between 1961 and 1979." Ph.D. diss., International University, 1981. 126 pp.

Describes the relationship between technological transfer and economic and job development. Emphasizes the transfer of agricultural machinery and fertilizer technology in relation to wheat and rice farming from 1961 to 1979.

1122. Anthony, Constance G. **Mechanization and Maize: Agriculture and the Politics of Technology Transfer in East Africa**. New York: Columbia University Press, 1988. 178 pp.

Discusses the influence of politics on technological development, diffusion and adoption as well as the effect of technology on the institutions that develop and introduce it. Describes the interaction of state and international organizations engaged in joint projects. Includes a discussion of the Green Revolution. Emphasizes Kenya and Tanzania.

1123. Arnon, I. **Agricultural Research and Technology Transfer**. Barking, United Kingdom: Elsevier, 1989. 841 pp.

This study suggests a detailed organizational plan to facilitate the transfer of agricultural technology to developing countries. Stresses the importance of research and extension systems.

1124. Barber, L.H. "Technological Transference? The Australian-New Zealand Farming Nexus in the Nineteenth Century." **Agricultural History** 57 (April 1983): 212-222.

New Zealanders imported sheep and shearing technology from Australia by the mid-nineteenth century. When wool prices dropped in the late nineteenth century New Zealanders adopted Australian refrigeration technology and diversified the sheep industry to export meat as well as wool.

1125. Burch, David. **Overseas Aid and the Transfer of Technology: The Political Economy of Agricultural Mechanization in the Third World**. Brookfield: Gower Publishing Co., 1987. 370 pp.

Argues that technological transfer depends more on the donor nations than on the needs of the recipient, developing nations. As a result, tractor and fertilizer production reached high levels but exports did not solve Third World problems. British technological transfer to Sri Lanka is an exception. It created a capital-intensive, productive agriculture.

1126. Compton, J. Lin, ed. **The Transformation of International Agricultural Research and Development**. Boulder: L. Rienner Publishers, 1989. 237 pp.

Evaluates the influence of the agriculture of the United States on developing nations, particularly in relation to the land-grant university and experiment station and extension systems. Discusses past policies and the role of women in programs for technological transfer.

1127. Deuson, R.R., and J.C. Day. "Transfer of Sustainable Technology in Dryland Agriculture: Lessons from the Sahel in the 1980s." **Agricultural Economics** 4 no. 3/4 (1990): 255-66.

Argues that technologies as well as institutional and economic policies must be adapted to the harsh, dry environment of the Sahel, if farmers are to become more productive.

1128. Duvick, D.N. "Research Collaboration and Technological Transfer: The Public and Private Sectors in Developing Countries and the International Seed Companies." In **Strengthening Collaboration in Biotechnology: International Agricultural Research and the Private Sector. Proceedings of a Conference Held April 17-21, 1988 in Rosslyn, Virginia**, edited by J.I. Cohen, 21-32, 1989. Washington, D.C.: Agency for International Development, 1989.

Urges cooperation via shared research grants. Contends the public international agricultural research centers must encourage and help integrate biotechnology in the plant breeding endeavors of developing nations. Also urges the private international seed companies to share their breeding and research programs with nations that have inadequate agricultural production.

1129. Herog, V.A. "Transfert de Connaissances en Afrique: Histoire et Enseignements pour le Futur." **African Development** 5 no. 1 (1980): 39-66.

Examines technological transfer among African nations. Includes a discussion of the introduction of maize and manioc to Africa and patterns of technological diffusion.

1130. Kerr, W.A. "Technological Transfer Through Pure-bred Herds in British Columbia." **Agricultural History** 65 (Winter 1991): 72-99.

Discusses the importance of land and commercial cattle companies and the gentry class in transferring breeding improvements to small-scale livestock producers. Uses Shorthorns and Herefords as a case study.

1131. Pineiro, M.E. "Generation and Transfer of Technology for Poor Small Farmers. In **Technology Systems for Small Farmers: Issues and Options**, edited by A.M. Kesseba, 45-67. Boulder: Westview Press, 1989.

Contends that technological transfer has failed in Latin America because small-scale farmers do not have the capital to invest in new equipment, because farmers are tied to the land and are unwilling to be displaced and because of an unwillingness to change. Urges a systems approach in which research, technology and social and economic considerations are integrated.

1132. Rerkasem, K., and B. Rerkasem. "Technology Transfer and Adoption in Irrigated Agriculture: A Case Study of Mae Kung Village in the Chang Mai Valley." In **Agricultural Information and Technological Change in Northern Thailand**, edited by M. Kaosa-Ard, K. Rerkasem and C. Roongruangsee, 111-31. N.P.: Thailand Development Research Institute, 1989.

Discusses the dissemination of irrigation technology in the Mae Kung village in Thailand's Sao Pa Tong District. Stresses the importance of the village information network and the knowledge of farmers in the adoption of new technology.

1133. Ruttan, V.W., and Yujiro Hayami. "Technology Transfer and Agricultural Development." **Technology and Culture** 14 no. 2 (1973): 119-151.

Compares technological transfer in Russia and Japan as well as among the nations involved with the Green Revolution. Concludes that technological transfer is most successful in the developing nations that have institutional support or access to an international research network.

1134. Ryan, Bryce. "A Study in Technological Diffusion." **Rural Sociology** 13 (September 1948): 273-85.

A case study of the diffusion of hybrid corn among farmers in Iowa. Shows farmers adopting hybrid seed corn in small quantities at first and a relatively long period of slow growth followed by rapid acceptance. The trial run process and low cost enabled farmers to make up their own minds based on personal experience

1135. Sims, H., and D. Leonard. "The Political Economy of the Development and Transfer of Agricultural Technologies." **Linkages Theme Paper, International Service for National Agricultural Research No. 4** (1989): 24 pp.

Discusses the factors that inhibit the development of technology that can be transferred to another culture including social and economic problems. Theoretical in methodology and argument.

1136. Ziemke, M.C. "Technological Transfer from the Dim Past: Alternatives to Reinventing the Wheel." **Journal of Technology Transfer** 14 (Spring 1989): 41-43.

Brief discussion of the history of technological transfer that includes a note on internal combustion engines.

CHAPTER XV

BIOTECHNOLOGY

1137. Atthasampunna, P., M. Suwana-adth, and A. Bhumiratana. "Current Biotechnological Developments in Thailand." **Critical Reviews in Biotechnology** 9 no. 1 (1989): 41-59.

Reviews the history of biotechnology in Thailand, includes reference to plant breeding and pathology as well as animal breeding and the preservation of germplasm. Notes the work of the Regional Network for Microbiology in Southeast Asia and the development of national policy.

1138. "Biotechnologia: Perspectiva General y Desarrollos en America Latina." **Progreso Economico y Social en America Latina**, 207-302. Washington, D.C.: Interamerican Development Bank, 1988.

An introductory history of biotechnology with emphasis on its relationship to agriculture in Latin America. Topics include tissue culture for the regeneration of plants and the introduction of new crops for semi-tropical and tropical environments. Contends that biotechnology means a simpler and more appropriate technology for solving agricultural problems.

1139. Bonjean, Alain. **Le Ricin: Une Culture pour la Chimie Fine**. Paris: Galileo, 1991. 101 pp.

A biotechnological study of the cultivation of castor beans for the extraction of chemicals for agricultural purposes.

1140. Bonny, S., and J.P. Roubaud, ed. "Les Nouvelles Technologies: Quels Impacts sur L'Agriculture et L'Agro-Alimentaire? Colloque des 21 et 22 Septembre 1988." **Economie Rurale** no. 192-193 (1989): 152 pp.

A discussion of new technologies that affect agriculture and the food industries. Emphasizes animal genetics, robotics and ethical problems. The technologies discussed may involve changes in land use, employment and international relationships.

1141. Boonman, J.G. "De Gentechnologie en het Kweekbedrijf." **Prophyta** [Netherlands] 42 no. 9 (1988): 214-16.

A study of gene technology and plant breeding in the Netherlands since 1973. Concludes that plant breeders and geneticists need greater contact and cooperation.

1142. Botterman, J. "Improvement of Agricultural Crops by Genetic Engineering." **Mededelingen van de Faculteit Landbouwwetenschappen, Rijksuniversiteit Gent** [Netherlands] 53 no. 4a (1988): 1695-99.

A brief historical overview of biotechnology in relation to crop improvement. Discusses breeding for resistance to herbicides, insects and viruses.

1143. Boxus, P. "La Multiplication in Vitro, une Biotechnologie Interessant pour le Development: Ses Perspectives Industrielles." **Annales de Gembloux** [Belgium] 95 no. 3 (1989): 163-81.

Discusses the history of in vitro propagation and its prospects for commercial use, especially in relation to the production of potatoes, ornamentals and fruits.

1144. _____. "La Maitrise des Techniques de Multipliction in Vitro: Realities et Perspectives." **Annales de Gembloux** [Belgium] 96 no. 1 (1990): 33-42.

A historical sketch of the research on the in vitro propagation of plants. Discusses techniques and research in thirteen West European countries.

1145. Bud, Robert. **Biotechnology in the Twentieth Century**. London: Science Museum, 1989. 73 pp.

Traces the first links between biology and technology to Denmark, Germany and Hungary prior to World War I.

Scientists and engineers in Great Britain, Sweden and the United States spread this knowledge in their publications. Discusses the development of biotechnology as a scientific discipline.

1146. Core, J. "Is There Truly a Need for Biotechnology in the Dairy Sector and Is There a Price to be Paid for Its Use? A Producer Perspective." **Proceedings of the XIII International Dairy Congress, Montreal, October 8-12, 1990, Vol. 1, 722-76.** Brussels: International Dairy Federation, 1991.

Although dairymen always have been quick to adopt new technologies, the application of biotechnology to help increase production has met some resistance because it affects the social as well as the economic aspects of dairying.

1147. Deo, S.D., and N. Mohseni. "Biotechnology and Development of Agriculture in Third World Countries: The Next Phase of Technology Transfer." In **Sociology of Agriculture: Technology, Labour, Development and Social Classes in an International Perspective**, edited by A. Bonanno, 27-54. New Delhi: Concept Publishing Co., 1989.

Argues that new developments in biotechnology do not meet the needs of small-scale farmers with little capital in developing nations. Rather, it will increase concentrations of power and further contribute to displacement. Includes a general discussion of technological transfer in developing nations.

1148. Gerhartz, W. **Enzymes in Industry: Production and Applications**. Weinheim, Federal Republic of Germany: VCH Verlagsgelleschaft, 1990. 321 pp.

Discusses the use of enzymes for the processing of meat, dairy products, fruits, vegetables and wine. Notes the use of enzymes in genetic engineering as well as environmental concerns.

1149. Goodman, D., and H. Buller. "Farming and Biotechnology: New Approaches to Rural Development." In **Rural Development: Problems and Practices**, edited by H. Buller and S. Wright, 97-108. Aldershot, United kingdom: Avebury, 1990.

Discusses the potential of biotechnology in relation to the fermentation of feeds and the genetic engineering of crops to produce custom-made foods.

1150. Groosman, T., A. Linnemann, and H. Wierema. "Seed Industry in Kenya." In **Seed Industry Development in a North/South Perspective**, 39-61, Wageningen, Netherlands: Centre for Agricultural Publishing and Documentation, 1991.

An overview of Kenyan agriculture with an emphasis on the development of new varieties. Notes the history of the seed industry in relation to maize, wheat, barley, rice and sunflowers. Contends the biotechnological applications of the international seed companies will change the Kenyan seed industry. Urges research to capitalize on Kenya's large genetic plant base.

1151. Kloppenburg, Jack R., Jr. **First the Seed: The Political Economy of Plant Biotechnology, 1492-2000**. Cambridge: Cambridge University Press, 1988. 349 pp.

A history of scientific and commercial plant breeding in the United States. Contends that hybridization has given the private seed companies control of germplasm and markets at the expense of individual breeders and farmers.

1152. Kung, Shain-dow, and Charles J. Arntz, eds. **Plant Biotechnology** Boston: Butterworths, 1989. 423 pp.

Locates the foundation of biotechnology in the efforts of agriculturists to improve plants and animals by selection and cross breeding. Reviews the major topics relevant to plant biotechnology. Historians will find the sections on techniques and practical uses to be the most helpful.

1153. MacDonald, June Fessenden, ed. **Agricultural Biotechnology at the Crossroads: Biological, Social & Institutional Concerns**. Ithaca, NY: National Agricultural Biotechnology Council, 1991. 307 pp.

See for a discussion of biological pest control problems, including ethical considerations.

1154. Micke, A., B. Donini, and M. Maluszynski. "Induced Mutation for Crop Improvement." **Mutation Breeding Review** no. 7 (1990): 41 pp.

Surveys the history of plant mutation for crop improvement. Relates mutation breeding to traditional plant breeding. Lists varieties obtained.

1155. Monteville, T.J. "The Evolving Impact of Biotechnology on Food Microbiology." **Journal of Food Safety** 10 no. 2 (1990): 87-97.

Discusses the application of biotechnology to food technology. Useful for a contemporary perspective and historical background.

1156. Prave, P., et al., ed. **Fundamentals of Biotechnology**. Weinheim, Federal Republic of Germany: VCH Verlagsgesselschaft, 1987. 792 pp.

Includes a useful section on the history of biotechnology as well as environmental biotechnology.

1157. Reeve, J.E. "The Use of Hormones in Animal Rearing." **Proceedings of the Nutrition Society of New Zealand**, 54-58, 1988.

A historical sketch of growth enhancers and their use in Europe. Notes possible health hazards in the consumption of meat. Contends that natural hormones, such as oestrogens, progestogens and androgens, do not endanger human health.

1158. Russell, A.M. **The Biotechnology Revolution: An International Perspective**. Brighton, United Kingdom: Wheatsheaf Books, 1988. 266 pp.

Discusses the political and economic significance of genetic engineering in an international perspective. Urges international cooperation to help eliminate the risks as well as to share the benefits of biotechnological changes in agriculture.

1159. Ruttan, V.W. "Productivity Implications of Biotechnology." In **Genetic Improvement of Agriculturally Important Crops: Progress and Issues**, edited by Robert T. Fraley, Nicholas M. Frey and Jeff Schell, 107-12. Cold Spring Harbor, NY: Cold Spring Harbor Laboratory, 1988.

A historical note on plant biotechnology and agricultural policy in the United States.

1160. Schlee, D., and H.P. Kleber. **Worterbuch der Biologie. Biotechnologie**. Jena, German Democratic Republic: Gustav Fisher Verlag, 1991. 1,096 pp.

This dictionary of biotechnology contains more than 4,000 terms in German. Historians as well as scientists will find it useful.

1161. Schmauder, H.P., and P. Doebel. "Plant Cell Cultivation as a Biotechnological Method." **Acta Biotechnologica** 10 no. 6 (1990): 501-16.

A brief history of plant biotechnology regarding propagation, breeding and in vitro cultivation. Includes some discussion of the economics.

1162. Schon, H. "Formen un Entwicklung des Technischen Fortschritts in der Landwirtschaft." **Politischen Studien** 39 no. 301 (1988): 549-70.

A critique of technological change in German agriculture. Includes a discussion of biotechnology as well as hardware developments, such as the tractor.

1163. Sundquist, W.B. "Emerging Maize Biotechnologies and their Potential Impact." **Technical Papers--OECD Development Centre, No. 8**, 1989. 29 pp.

Discusses the new biotechnologies for the production of maize, particularly in relation to improving the quality of the starch, protein and oil as well as developing plants that will resist herbicides, pests and bad weather. Notes environmental concerns and the need to maintain traditional plant breeding programs.

1164. Theret, M. "Evolution et Revolution Scientifiques et Techniques en Elevage et dans ses Productions Depuis 1750." **Comptes Rendus de l'Academie d'Agriculture de France** 74 no. 6 (1988): 47-56.

A brief history of animal breeding in France since 1830. Includes crossbreeding and the use of biotechnology.

1165. Whitaker, John R. "Interdependence of Enzymology and Agricultural Biotechnology." **ACS Symposium Series No. 389**, 1-9. Washington, D.C., 1989.

A brief historical survey of the relationship of enzymology to agricultural biotechnology. Notes the discovery and application of enzymes by agricultural scientists and discusses the current status of enzymology.

CHAPTER XVI

THE GREEN REVOLUTION

1166. Alauddin, M., and C.A. Tisdell. "The 'Green Revolution' and Labor Absorption in Bangladesh Agriculture: Relevance of East Asian Experience." **Research Paper No. 229**, University of Melbourne, 1989. 20 pp.

    Contends the effect of the Green Revolution on Japan and Taiwan will not apply to Bangladesh. Instead of reducing the number of agricultural laborers, the authors argue that agricultural labor will become more intensified.

1167. Bickel, Leonard. **Facing Starvation: Norman Borlaug and the Fight Against Hunger**. New York: Reader's Digest Press, 1974. 376 pp.

    A sympathetic biography designed for the general reader rather than the historian of agricultural science.

1168. Borlaug, Norman. "The Green Revolution: The Role of CIMMYT and Wheat Lies Ahead." In **Trade and Development. Proceedings of the Winter 1986 Meeting of the International Agricultural Trade Research Consortium**, edited by M.D. Shane, 113-127. United States Department of Agriculture, Economic Research Service, 1988.

A review of the major achievements of the Centro Internacional de Mejoramiento de Maiz y Trigo (CIMMYT) during the previous twenty years. Emphasizes genetic improvements in maize, the diffusion of semidwarf wheat varieties and crop production research, especially for wheat and corn, to help farmers in developing nations.

1169. Brown, Lester R. "The Agricultural Revolution in Asia." **Foreign Affairs** 46 (July 1968): 688-98.

An important supplement to the scientific literature on the Green Revolution in Asia. Stresses the importance of political commitment as well as the introduction of high-yielding varieties of cereal grains. Notes the importance of technical, economic and social adjustments necessary to accommodate the Green Revolution.

1170. Byres, T.J. "The Dialectic of India's Green Revolution." **South Asian Review** [Great Britain] 5 no. 2 (1972): 99-116.

Analyzes the Green Revolution in India in regard to increased productivity of wheat and a decline in agricultural employment. Shows the Green Revolution does not bring unmitigated progress.

1171. Chakravarti, A.K. "Green Revolution in India." **Annals of the Association of American Geographers** 63 no. 3 (1973): 319-336.

Argues that the Green Revolution has not solved the problem of famine in India. Even though hybrid seeds have doubled the agricultural production of food grains, the High Yielding Variety Seed Program (HVP) caused serious civil disputes during the early 1970s.

1172. Chennamaneni, Ramesh. "Zu den Ergebnissen der 'Grun Revolution' in Indien Anhand Soziologisch-Okonomischer Untersuchungen in Einigen Dorfern Andhra Pradeshs." **Asien, Afrika, Lateinamerika** [East Germany] 17 no.1 (1989): 43-57.

Reviews the implementation of the Green Revolution in India from 1965 to 1985. Supplemented with interviews, conducted in 1985, of individuals in four villages of Karimnager, Andhra Pradesh.

1173. Dagognet, F. **Des Revolutions Vertes. Histoire et Principes de l'Agronomie**. Paris: Lyon University, 1973. 184 pp.

Discusses the influence of science, agronomy and agro-biology on the agrarian revolutions in France between the eighteenth and the late nineteenth centuries. Contends that new skills for the manipulation of science and technology will become more important than the right to own land.

1174. Dahlberg, K.A. "The Industrial Model and Impacts on Small Farmers: The Green Revolution as a Case." In **Agroecology and Small Farm Development**, edited by M.A. Altieri and S.B. Hecht, 83-90. Boca Raton, FL: CRC Press, 1990.

Argues the key features of the Green Revolution are increased production due to science and a belief in the neutrality of technology. Also discusses the social, political and economic aspects of the Green Revolution. Contends that an industrial approach to agriculture has negative consequences for small-scale farmers.

1175. Dalrymple, Dana G. "Development and Spread of High-Yielding Varieties of Wheat and Rice in the Less Developed Nations." United States Department of Agriculture, **Foreign Agricultural Economic Report, No. 95**, 1974. 120 pp.

Reviews the development of high-yielding wheat and rice. Emphasizes semidwarf wheat varieties developed at the International Maize and Wheat Improvement Center in Mexico and rice varieties developed in the Philippines at the International Rice Research Institute. Covers the period from 1965 to 1974.

1176. _____. "Measuring the Green Revolution: The Impact of Research on Wheat and Rice Production." United States Department of Agriculture, **Foreign Agricultural Economic Report, No. 106**, 1975. 40 pp.

Reviews the major considerations for evaluating the effects of international research programs on crop production in the Third World, particularly for high-yielding varieties of wheat and rice.

1177. Farmer, B.H. "Perspectives on the 'Green Revolution' in South Asia." **Modern Asian Studies** [Great Britain] 20 no. 1 (1986): 175-199.

Reviews the effects of the Green Revolution on agricultural production in India, Pakistan, Bangladesh and Sri Lanka.

1178. Germain, R. "La Revolution Verte: Ses Origines, Ses Succes, Ses Contraintes." **Bulletin des Seances de l'Academie des l'Outre-Mer** [Belgium] 25 no.4 (1979): 649-662.

Looks at the beginning of the Green Revolution in Mexico during the mid-1940s, where yields increased for corn and wheat, and in Asia, where rice yields were improved. In both regions, however, failures caused by insufficient irrigation, inadequate pest control and resistance to innovation limited the success of the Green Revolution.

1179. Gough, Kathleen. "The Green Revolution in South India and North Vietnam." **Monthly Review** 29 no. 8 (1978): 10-21.

Comparison of rice growing regions of the Thanjavur district in southeast India and the Thi Binh province of North Vietnam. Argues that socialist land reform since 1965 in Vietnam has enabled farmers to out produce their Indian counterparts even though both use high yield varieties, new technology and pesticides.

1180. Greenland, D.J. "Bringing the Green Revolution to the Shifting Cultivator." **Science** 190 (November 28, 1975): 841-844.

A brief, general discussion of the Green Revolution's effect on agriculture in the nonwestern world.

1181. Griffin, Keith. **The Political Economy of Agrarian Change**. Cambridge: Harvard University Press, 1974. 264 pp.

Discusses the effects of the Green Revolution in Asia, Mexico, Columbia, India, Pakistan and the Philippines. Argues the Green Revolution has benefited the more prosperous regions and farmers and that it has not solved the food problems in the developing nations because of population increases and the loss of arable lands.

1182. Gurvich, R.P. "'Zelenaia Revolutsiia' v Indi: Sotsial' No-Ekonomicheskie Rezul' Taty." **Norody Azii i Afriki** [Soviet Union] no. 1 (1971): 17-28.

Surveys the effects of the Green Revolution in India. Contends that it has primarily benefited the wealthy farmers rather than helped solve the problems of rural poverty.

1183. Hayami, Yujiro. "Elements of Induced Innovation: A Historical Perspective for the Green Revolution." **Explorations in Economic History** 8 no. 4 (1971): 445-472.

Discusses the old rice economies of Japan, Korea and Taiwan and analyzes the transfer of technology and high-yield varieties of rice that created a Green Revolution in those countries between 1895 and 1969.

1184. Johnson, Stanley. "Food, People and Technology: A Report on the Green Revolution." **Vista** 7 no. 6 (1972): 28-32, 57.

A review of India's increased grain production during the 1960's due to the introduction of high-yield seeds and mechanization.

1185. _____. **The Green Revolution**. New York: Harper Torchbooks, 1972. 349 pp.

A sympathetic history of the Green Revolution for the general reader. Johnson, however, acknowledges the social, technical and economic problems that must be overcome before the scientific and technological advances that made the Green Revolution possible can reach fruition.

1186. Khan, S.A. **The State and Village Society: The Political Economy of Agricultural Development in Bangladesh**. Dhaka: University Press, 1989. 189 pp.

Contends the Green Revolution helped contain social and economic differences by spreading the benefits of increased agricultural production to many people. Even though large-scale landowners benefited the most from the Green Revolution, small-scale holders also improved their operations. Still, the Green Revolution has not generated sufficient capital to pay investments.

1187. Kislev, Yoav, and Michael Hoffman. "Research and Productivity in Wheat in Israel." **Journal of Development Studies** [Great Britain] 14 no. 2 (1981): 166-181.

Overview of the Green Revolution in Israel since the 1950's. Stresses the importance of wheat production.

1188. Kovalyov, E. "The Green Revolution: Its Technical and Social Aspects." **Internal Affairs** [Soviet Union] no. 9 (1972): 40-44.

Review of the technological aspects of the Green Revolution in Africa, Asia, and Latin America during the late 1960's.

1189. Ladejinsky, Wolf. "Ironies of India's Green Revolution." **Foreign Affairs** 48 (July 1970): 758-68.

Discusses the political and social problems caused by the Green Revolution, such as the polarization of income, increased rents for tenants and rural unemployment. Contends the success of the Green Revolution depends on the rapid growth of the entire economy and a high rate of employment.

1190. Nulty, Leslie. **The Green Revolution in West Pakistan: Implications of Technological Change**. New York: Praeger Publishers, 1972. 150 pp.

This study analyzes the development of agriculture in Pakistan from 1948 to 1970. Emphasis is placed on technological change. Contends the success of the Green Revolution in Pakistan was unique because the farmers in that area already had access to irrigation and chemical fertilizers before they adopted hybrid seeds.

1191. Parayil, Govindan. "Conceptualizing Technological Change: Technology Transfer in the Green Revolution." Ph.D. diss., Virginia Polytechnic Institute and State University, 1990. 257 pp.

Discusses economic, historical and sociological models for technological change and transfer. Uses the Green Revolution to test these theoretical models. This study is best avoided until the basic historical development of the Green Revolution has been mastered.

1192. _____. "The Green Revolution in India: A Case Study of Technological Change." **Technology and Culture** 33 (October 1992): 737-56.

A general history of technological transfer in India. Contends the Green Revolution required the education of peasant-farmers and that despite problems, such as the overuse of pesticides and machinery, it enabled India to become self-sufficient in food.

1193. Perkins, J.H. "The Rockefeller Foundation and the Green Revolution, 1941-1956." **Agriculture and Human Values** 7 (Summer/Fall 1990): 6-18.

A useful introduction to the work of the Foundation, particularly in relation to the development of agricultural science policy and the effect of the Green Revolution on Mexico and India.

1194. Powelson, John P. "Green Revolution in the Philippines." **Cultures et Development. Review International des Science du Developpement** [Belgium] 11 no. 4 (1979): 671-674.

Examines the Green Revolution in regard to production and profit and concludes that it failed to help the poor farmers in the Philippines during the 1970's.

1195. Pray, Carl E. "The Green Revolution as a Case Study in Transfer of Technology." **Annals of the American Academy of Political and Social Science** no. 458 (1981): 68-80.

Contends that although the Green Revolution has not ended world hunger or poverty, it has dramatically increased the supply of cereal grains and kept food prices low.

1196. Romeiro, A.R. "Alternative Developments in Brazil." In **The Green Revolution Revisited: Critique and Alternatives**, edited by Bernhard Glaeser, 79-110. London: Allen & Unwin, 1987.

Discusses the influence of the Green Revolution on landownership, agricultural productivity, food prices, income distribution and technological change.

1197. Rudra, Ashok. "The Green and Greedy Revolution." **South Asian Review** [Great Britain] 4 no.4 (1971): 291-305.

Argues that the Green Revolution has contributed to the inequitable distribution of wealth in India, because it is based on capitalist economic principles.

1198. Ruttan, Vernon W. "The Green Revolution: Seven Generalizations." **International Development Review** 19 no. 44 (1977): 16-23.

Discussion of the Green Revolution in Asia since the mid-1960's. Contends the new seed and fertilizer technology used to increase food grain production has delayed the revolutionary political and economic changes that were expected in rural areas by the mid-1970's.

1199. Sale, Kirkpatrick. **The Green Revolution: The Environmental Movement, 1962-1992**. New York: Hill & Wang, 1993. 128 pp.

Title misleading. A study of the origin and development of national environmental organizations and an assessment of their successes and failures.

1200. Shiva, Vandana. **The Violence of the Green Revolution: Ecological Degradation and Political Conflict in Punjab**. Dehra Dun, India: Research Foundation for Science and Ecology, 1989. 160 pp.

Discusses the ecological and political problems caused by the Green Revolution.

1201. Singh, Inderjit, and Richard H. Day. "A Microeconomic Chronicle of the Green Revolution." **Economic Development and Cultural Change** 23 no. 4 (1975): 661-686.

An early case study of the positive effects of the Green Revolution on farmers in the Punjab (India), particularly for technological change, production and employment. Based on data collected from the early 1950s until the mid-1960s.

1202. Stakman, Elvin C., Richard Bradfield, and Paul C. Mangelsdorf. **Campaigns Against Hunger**. Cambridge: Harvard University Press, 1967. 328 pp.

Discusses the Rockefeller Foundation's work to increase food production, particularly in Mexico.

1203. Stone, Bruce. "Developments in Agricultural Technology." **China Quarterly** [Great Britain] no. 116 (December 1988): 767-822.

The Green Revolution in China is dependent on improved irrigation, availability of fertilizers and supplies of high-yield seed varieties. Although improved water control and access to improved varieties occurred during the Maoist era, fertilizer production lagged until the post-Mao period.

1204. Valarche, Jean. "La 'Revolution Verte' en Inde et l'Ideologie des Limites de la Croissance." **Civitas** [Switzerland] 29 nos. 1/2 (1973): 70-85.

Argues the Green Revolution greatly improved food supplies in India and enabled the government to address social problems, such as rapid population growth, without worrying about agricultural production.

1205. Vega, Belita Amihan. "The Structure of Knowledge Production for Philippine Agriculture." Ph.D. diss., University of Wisconsin, 1990. 640 pp.

A study of the institutional framework for agricultural research in the Philippines. Includes a discussion of the Green Revolution. Contends that agricultural knowledge is not only a technical matter but also a political and economic activity. Argues that the underdevelopment of agriculture in the Philippines is due to the structure of institutional research.

1206. Venkateswarlu, B. **Dynamics of Green Revolution in India**. New Delhi: Agricole Public Academy, 1985. 309 pp.

Analyzes the Green Revolution in India from 1950 to 1980. Discusses crop raising and production trends, technological adoption and the envelopment of research and extension.

1207. Wade, Nicholas. "Green Revolution: Creators Still Quite Hopeful on World Food." **Science** 185 (September 6, 1974): 844-845.

Rebuttal to the critics of the Green Revolution. Acknowledges the continued problem of the population explosion and the difficulties of high-yield agriculture to meet increasing demands for food.

1208. _____. "Green Revolution (I): A Just Technology, Often Unjust in Use." **Science** 186 (December 27, 1974): 1093-1096.

Presents the negative side of the Green Revolution. Contends the Third World has been unable to adopt Western-style agriculture and that the Green Revolution has failed.

1209. Wharton, Clifton R., Jr. "The Green Revolution: Cornucopia or Pandora's Box?" **Foreign Affairs** 47 (April 1969): 464-76.

A good review of the positive and negative effects of the Green

Revolution. Essential reading for all research on this topic.

CHAPTER XVII

NUTRITION

1210. Beatty, W.K. "The History of Nutrition: A Tour of the Literature." **Federation Proceedings, Federation of American Societies for Experimental Biology** 36 (October 1977): 2511-13.

A listing of the basic sources on the history of nutrition, and a useful introduction to the subject.

1211. Becker, Stanley L. "Butter Makes Them Grow: An Episode in the Discovery of Vitamins." Connecticut Agricultural Experiment Station, **Bulletin, No. 767**. 1977. 21 pp.

Discusses the work of Thomas B. Osborne and Lafayette B. Mendel, who discovered the importance of vitamins, amino acids and proteins for growth and diet. Emphasizes their experiments from 1909 to 1917.

1212. Bennett, Merritt K. and R.H. Pierce. "Change in the American National Diet, 1879-1959," 95-119. **Food Research Institute Studies, No. 2**. Stanford: Stanford University, 1961.

Examines the major trends in the American diet over eight decades. Includes statistical data on consumption.

1213. Bing, F.C. "Dietary Fiber in Historical Perspective." **Journal of the American Dietetic Association** 69 (November 1976): 498-505.

A succinct review of the research on the relationship between low consumption of dietary fiber and disease. Notes suggestions for further research.

1214. _____. "Nutrition Research and Education in the Age of Franklin." **Journal of the American Dietetic Association** 68 (January 1976): 14-21.

Discusses Benjamin Franklin and William Start in relation to their interests in nutrition during the eighteenth century.

1215. Bing, F.C., and Harry J. Prebluda. "E.V. McCollum: Pathfinder in Nutrition Investigations and World Agriculture." **Agricultural History** 54 (January 1980): 157-66.

Between 1903 and 1967, E.V. McCollum helped discover and analyze several vitamins, and he worked to improve animal feeds. McCollum's research emphasized dairy products, and he contributed to the development of the nonfat dried milk industry.

1216. Briggs, G.M. "Why Study Nutrition History?" **Federation Proceedings, Federation of American Societies for Experimental Biology** 36 (May 1977): 1905.

A note on rationale, including the importance of context, nomenclature, biography and culture. This is an introduction to papers presented at a symposium on "Selected Topics in History of Nutrition" during the 60th Annual Meeting of the Federation of American Societies for Experimental Biology on April 12, 1976.

1217. Brown, Myrtle L., ed. **Present Knowledge in Nutrition**. 6th ed. Washington, D.C.: International Life Sciences Institute-Nutrition Foundations, 1990. 532 pp.

Discusses the current status of nutrition research. Topics include degenerative diseases, brain development, immunity and toxicants. Notes historical landmarks in nutrition, and includes an essential bibliography.

1218. Cahill, G.F. Jr. "The Future of Carbohydrates in Human Nutrition." **Nutrition Review** 44 (February 1986): 40-43.

Surveys the history of carbohydrates in human food from prehistory to the present.

1219. Carpenter, K.J. "The History of Enthusiasm for Protein." **Journal of Nutrition** 116 (July 1986): 1364-70.

Traces the history of protein requirements from the late nineteenth century to the present. Notes a sharp decline in daily recommended allowances. Discusses current thought about the relationships between protein and calories in daily diet.

1220. Darby, W.J. "Nutrition: Gastronomy, Mythology or Science." **Nutrition Today** 21 (September/October 1986): 4-11.

Discuses the history of therapeutic diets, the value of vitamin C, nutrition for children, nutrition faddism and quackery. Surveys food cults in the United States and the work of scientists to disseminate information about good nutrition.

1221. _____. "Nutrition Science: An Overview of American Genius." United States Agricultural Research Service, W.O. Atwater Memorial Lecture, Washington, D.C., 1976. 39 pp.

Discusses two centuries of nutritional science. Notes the contributions of individuals, government, media, industry and organizations.

1222. _____. "The Science of Nutrition, 1776-1976." **Utah Science** 37 (September 1976): 70-74.

Dates the foundation of nutrition science to the discovery of oxygen and oxidation during the late eighteenth century. Focuses on the work of Lavoisier, Priestly and Scheele, and describes the influence of Justus von Liebig, W.O. Atwater, Graham Lusk and Casimio Funk on the history of nutrition.

1223. Etheridge, Elizabeth W. **The Butterfly Caste: A Social History of Pellagra in the South**. Westport, CT: Greenwood Publishing Co., 1972. 278 pp.

Discusses this illness created by nutritional deficiencies in the American South. Describes the work of Joseph Golden and his discovery of the cause of pellagra.

1224. Friend, Berta. "Nutrients in United States Food Supply: A Review of Trends, 1909-1913 to 1965." **American Journal of Clinical Nutrition** 20 (August 1967): 907-914.

Estimates the food energy in protein, fat, carbohydrates, calcium, iron and other vitamins and minerals in the U.S. during various periods. This study also provides the percentage of enrichments and fortifications in certain foods.

1225. Goldblith, S.A., and M.A. Joslyn. **Milestones in Nutrition**. 2 Vols. Westport, CT: Avi Publishing Co., 1964.

Traces the past 150 years of nutritional sciences. The topics include the early history of nutrition, the feeding of purified diets, vitamins and minerals and the essential fatty acids.

1226. Gordon, K.D. "Evolutionary Perspectives on Human Diet." In **Nutritional Anthropology**, edited by Francis E. Johnston, 3-39. New York: A.R. Liss, 1987.

Discusses the evolution of prehistoric diet. Evaluates the shift from vegetarianism to hunting and gathering to the domestication of plants and animals. Notes the problems of evaluating circumstantial evidence.

1227. Harper, A.E. "Nutrition: From Myth and Magic to Science." **Nutrition Today** 23 (February 1988): 8-17.

Traces the contributions of nutrition science to public health. Outlines the development of nutrition from prehistory to its development as a biological and applied medical science. Addresses the division in society regarding nutrition science and popular pseudoscientific nutritional practices.

1228. Harris, Leslie J. "The Discovery of Vitamins." In **The Chemistry of Life**, edited by Joseph Needham, 156-70. Cambridge: Cambridge University Press, 1970.

A history of vitamins that emphasizes the work of C. Eijkman, C. Funk and Sir F.G. Hopkins. Covers the period since 1912 regarding the discovery of vitamins A, B, and D.

1229. Hartog, C. den. "Artsen, Landbouw, Voedselvoorziening en Voeding in de Periode van 1850-1950." **Voeding** [Netherlands] 45 no. 4 (1984): 134-139.

During the nineteenth century with the application of science to agriculture, medical doctors often experimented with fertilization of the soil and the cure and prevention of livestock diseases. In the twentieth century doctors continued to influence rural life by advising farmers and other residents about matters of nutrition.

1230. Hegsted, D.M. "Nutrition: The Changing Scene." **Nutrition Review** 43 (December 1985): 357-367.

An excellent historical discussion of nutritional research, the discovery of the vitamins and the relationship of nutrition and chronic diseases.

1231. Holmes, Frederic L. "The Transformation of the Science of Nutrition." **Journal of the History of Biology** 8 (Spring 1975): 135-144.

Discusses the contributions of Albrecht von Haller, Hermann Boerhaave, Georges Buffon, Antoine Lavoisier, Georges Cuvier and Rene Dutrochet during the eighteenth and nineteenth centuries.

1232. Ihde, Aaron J. and Stanley Becker. "Conflict of Concepts in Early Vitamin Studies." **Journal of the History of Biology** 4 (Spring 1971): 1-33.

Discusses the major diseases caused by vitamin deficiencies during the eighteenth and nineteenth centuries and the research to discover the causes and cures. Notes the work of the agricultural experiment stations.

1233. Jaffe, W.G. and Bengoa, J.M. "Nutrition Ayer y Hoy." **Archivos Latinoaericanos de Nutricion** 38 no. 3 (1988): 429-44.

Outlines the history of nutrition from the prehistoric period to the present. Contends that agricultural changes disrupted the nutritional balance of hunter-gatherers. Notes changing food habits and the origin of nutrition science. Problems in human nutrition are related to changes in agriculture and food technology.

1234. Kanarek, R.B., and Gambill N. Orthen. "Complex Interactions Affecting Nutrition-Behavior Research." **Nutrition Review** 44 (May 1986): 172-175.

Notes the relationship between an individual's ideas about food and health and the influence of those ideas on nutrition research.

1235. King, Charles Glen. **A Good Idea: The History of the Nutrition Foundation**. New York: Nutrition Foundation, 1976. 241 pp.

Provides an overview from the creation of the Foundation in 1941. Discusses the major scientific research on nutrition in the areas of vitamins, metabolism, protein and amino acids. Includes a perspective on public policy.

1236. Kirkland, Edward C. "'Scientific Eating': New Englanders Prepare and Promote a Reform, 1873-1907." **Proceedings of the Massachusetts Historical Society** 86 (1974): 28-52.

Describes elitist nutritional reform and agricultural experimentation during the late nineteenth century, particularly the work of Wilbur Olin Atwater, Edward Atkinson and Ellen Swallow Richards. Notes the German influence on the social, economic and political aspects of reform.

1237. Leverton, R.M., et al. "Research in Agriculture and the Profession of Dietetics." **Journal of the American Dietetic Association** 64 (June 1974): 638-41.

Notes the most important research and publications in the areas of nutrition, dietary surveys, nutritive values and food preparation from the 1890s to the present. Includes the work of the USDA, state agricultural experiment stations and U.S. Public Health Service.

1238. Lusk, Graham. **Nutrition**. New York: P.B. Hoebner, 1933. 142 pp.

Traces the history of nutrition from antiquity to the early twentieth century. Designed for the layreader but scholars will find it a dated but useful introduction.

1239. McCay, C.M. **Notes on the History of Nutrition Research**. Berne: Hans Huber Publishers, 1973. 234 pp.

Discusses the work at the beginning of the nineteenth century to improve the transformation of feed into milk and meat. Notes the relationship of nitrogen in the air to plants, livestock and soil.

1240. McCollum, Elmer V. **A History of Nutrition: The Sequence of Ideas in Nutritional Investigations**. Boston: Houghton Mifflin Co., 1957. 451 pp.

Traces the work of the early clinicians who sought to discover the effects of foods on the sick and the well. Covers the period from the mid-eighteenth to the mid-nineteenth century.

1241. Maynard, Leonard A. "Early Days of Nutrition Research in the United States of America." **Nutrition Abstracts Review** 32 (April 1962): 345-55.

A brief survey of the most important pioneers in nutrition. Emphasizes the late nineteenth and twentieth centuries. Lists those scientists and their achievements in Europe and the United States.

1242. Munro, H.N. "Back to Basics: An Evolutionary Odyssey with Reflections on the Nutrition Research of Tomorrow." **Annual Review of Nutrition** 6 (1986): 1-12.

Stresses the importance of applying molecular and cell biology to nutrition science.

1243. Roberts, Lydia J. "Beginnings of the Recommended Dietary Allowances." **Journal of the American Dietetic Association** 34 (September 1958): 903-08.

Recounts the reasons for the establishment of Recommended Dietary Allowances by a special committee appointed by President Franklin Delano Roosevelt. Established in 1941, the RDA serves as a guideline for the intake of calories, protein, minerals and vitamins in daily diet.

1244. Roe, Daphne A. **Early Contributions of the 1890 Land Grant Colleges to Nutrition Research**. Bethesda, MD: American Institute of Nutrition, 1987. 17 pp.

Discusses the human nutrition research at the black land-grant colleges in the United States.

1245. _____. "History of Promotion of Vegetable Cereal Diets." **Journal of Nutrition** 116 (July 1986): 1355-63.

Traces the history of vegetarianism in relation to human health. Urges nutritionists to respect the scientific value of vegetarianism.

1246. Shank, F.R., and V.L. Wilkening. "Considerations for Food Fortification Policy." **Cereal Foods World** 31 (October 1986): 728-40.

A historical review of food fortification research and the effects of food fortification on diet and health. Includes a discussion of food policy.

1247. Smith, C. Earle, Jr. **Man and His Foods: Studies in the Ethnobotany of Nutrition: Contemporary, Primitive, and Prehistoric Non-European Diets**. Tuscaloosa, AL: University of Alabama Press, 1973. 131 pp.

A summary of European and American nutritional history, including various problems. Describes the importance of ethnobotany to the study of nutrition.

1248. Smith, J., and J.S. Turner. "A Perspective on the History and Use of the Recommended Dietary Allowances." **Currents, the Journal of Food, Nutrition & Health** 2 no. 1 (1986): 4-11.

Discusses the history of RDA and the influence of dietary recommendations, particularly in relation to specific populations and nutritional objectives. Notes the work of the National Research Council in establishing allowances.

1249. Swan, Patricia B. "A History of Nutrition Research in the United States with Emphasis on Agricultural Experiment Stations." In **Human Resources Research, 1887-1987: Proceedings**, edited by Ruth E. Deacon and Wallace E. Huffman, 27-40. Ames: College of Home Economics, Iowa State University, 1986.

Reviews the development of human nutrition research, including political, economic and social influences as well as the major scientists. Places the work of the state agricultural experiment stations in historical perspective.

1250. _____. "Formula-Funded Nutrition Research in 1862 Universities--An Historical Perspective." In **AIN Symposium Proceedings: Nutrition '87**, edited by Orville A. Levander, 4-8. Bethesda, MD: American Institute of Nutrition, 1987.

A brief overview of human nutrition research at the land-grant universities. Should be read along with the previous citation.

1251. Todhunter, Elizabeth Neige. "Chronology of Some Events in the Development and Application of the Science of Nutrition." **Nutrition Reviews** 34 (December 1976): 353-65.

A chronology of important occurrences and discoveries in the evolution of nutrition science. A good overview and a useful reference.

1252. _____. "Some Aspects of the History of Dietetics." **World Review of Nutrition and Dietetics** 18 (1973): 1-46.

A survey from primitive man to the present. Includes a topical discussion of various dietary diseases, the creation of dietary standards and the founding of the dietetic profession.

1253. _____. "The Story of Nutrition." In **Food: Yearbook of Agriculture, 1959**, 7-22. Washington, D.C.: United States Department of Agriculture, 1959.

Examines the progression of American nutrition as a science from the first work of W.O. Atwater in the nineteenth century to its status in the mid-twentieth century. Examines the work of the Ancient Greeks and European scientists as well as a host of American nutritional scientists, such as William Beaumont, Carl Voit, Max Rubner and Elmer V. McCollum.

1254. True, Alfred C., and R.D. Milner. "Development of the Nutrition Investigations of the Department of Agriculture." In **Yearbook of Agriculture, 1899**, 403-14. Washington, D.C.: United States Department of Agriculture, 1899.

Notes the research at the state experiment stations regarding the chemistry of food and the physiology of human nutrition. Includes some discussion of dietetic research for animals.

1255. Wadlinger, Mark H. **Bibliographical Survey of Vitamins, 1650-1930: With a Section on Patents**. Chicago: Mark H. Wadlinger, 1932. 334 pp.

Arranged chronologically. Separate entries are provided for each year. After 1915 the categories are divided by vitamins. Based on the collections in the John Crerar Library in Chicago. Dated but useful.

1256. Wadsworth, G. R. "The Use, Dietary Significance and Production of Fruit." **Journal of Human Nutrition** 32 (February 1978): 27-40.

Scientists have not yet determined the relationship between fruits in the human diet and the numerous physiological and biochemical processes of the body that determine the status of a person's health. Discusses the body processes after an individual has eaten fruit.

1257. Wilder, Russell M., and Robert R. Williams. "Enrichment of Flour and Bread: A History of the Movement." **National Research Council Bulletin, No. 110**, 1944. 130 pp.

Discusses the history of flour enrichment during the 1930s and 1940s. Includes the fixing of standards, the role of war agencies and state actions.

1258. Williams, Robert R. **Toward the Conquest of Beriberi**. Cambridge: Harvard University Press, 1961. 338 pp.

A discussion of the pathology of beriberi. Notes the practical application of flour enrichment research to rice milling in the Philippines.

1259. Wilson, Leonard. "The Clinical Definition of Scurvy and the Discovery of Vitamin C." **Journal of the History of Medicine and Allied Sciences** 30 (January 1975): 40-60.

Describes scurvy and the research to discover its cause and cure, particularly the work of Thomas Barlows, Axel Holst and Theodor Frolich. Includes a discussion of the discovery of vitamin C in 1919.

CHAPTER XVIII

INSTITUTIONS

1260. **A Decade of Learning: International Development Research Center, Agriculture, Food and Nutrition Sciences Division: The First Ten Years**. Ottawa, Canada: International Development Research Center, 1981. 180 pp.

See for Canadian agricultural assistance to underdeveloped areas of the world.

1261. Anderson, R.S., E. Levy, and B. Morrison. **Rice Science and Development Politics: Research Strategies and IRRI's Technology Confront Asian Diversity (1950-1980)**. Oxford: Clarendon Press, 1991. 394 pp.

A history of the founding and development of the International Rice Institute (IRRI) near Manila in 1960. The IRRI was designed to help solve Asian food problems. Emphasizes the IRRI's work in Sri Lanka and Bangladesh in relation to the Green Revolution.

1262. Anstey, T.H. **One Hundred Harvests: Research Branch, Agriculture Canada, 1886-1986**. Historical Series No. 27. Ottawa, Canada: Research Branch, Agriculture Canada, 1986. 423 pp.

Commemorates a century of agricultural research at Agriculture Canada, particularly in relation to supporting the food industry. Includes some discussion of its links with the Commonwealth Agricultural Bureau.

1263. _____. "The Formation of the Experiment Farms." **Prairie Forum** 11 (February 1986): 185-94.

In 1886, the Canadian Experimental Farm Station Act created a special bureau within the Department of Agriculture. This experimental farm bureau had the responsibility to conduct scientific experiments to improve Canadian agricultural practices.

1264. Baker, Gladys, et al. **Century of Service: The First One Hundred Years of the United States Department of Agriculture**. Washington, D.C.: United States Department of Agriculture, 1963. 560 pp.

The standard institutional history of the USDA. A detailed reference.

1265. _____. "The Agricultural Research Center of the United States Department of Agriculture." United States Department of Agriculture, **Information Bulletin No. 189**, 1958. 46 pp.

An organizational description of the Agricultural Research Center at Beltsville, Maryland.

1266. Bartels, R. "Ruckblick auf 110 Jahre Feldversuche im Bodentechnologischen Institut." **Zeitschrift fur Kulturtechnik und Landentwicklung** 31 no. 2 (1990): 67-71.

A history of the Institute of Soil Technology in the Federal Republic of Germany. During the Institute's century of existence, its research emphasis has changed from developing cultivation methods for bogs to conservation tillage measures designed to protect the environment in the lowlands.

1267. Bradfield, Richard. "The Sciences, Pure and Applied, in the First Century of the Land-Grant Institutions." **Science Education** 46 (April 1962): 240-47.

Discusses the most significant contributions of the land-grant colleges since the Morrill Act in 1862. Some discussion of the development of the Morrill Act and the people involved.

1268. Bourgeois, R. "Structural Linkages for Integrating Agricultural Research and Extension." **Working Paper, International Service for National Agricultural Research No. 35**. The Hague, Netherlands: CABI, 1990. 33 pp.

Discusses the problems encountered in disseminating agricultural research through extension systems. Contends that researchers and extension agents must work in partnership. Includes case studies.

1269. Brown, William L. "USDA Contributions to Progress in Plant Genetics." **Agricultural History** 64 (Spring 1990): 313-18.

A brief survey of the plant breeding activities in relation to disease resistance and genetics in the USDA. Includes some biographical information.

1270. _____, and William B. Lacy, ed. **The Agricultural Scientific Enterprise: A System in Transition**. Boulder: Westview Press, 1986. 348 pp.

A critical examination of publicly supported agricultural research in the United States. Emphasizes the work of the state agricultural experiment stations.

1271. Clayton, O. "The History of the Canadian Seed Growers' Association." **Plant Varieties and Seeds** 3 no. 3 (1990): 127-38.

Established in 1904, the Canadian Seed Growers' Association (CSGA) has established a close working relationship with the Canadian government. Since 1959, it has been the only legally recognized "pedigreeing" or "certifying" seed organization in Canada. It has promoted research, plant breeding and technological transfer.

1272. Compton, J.L., ed. **Transformation of International Agricultural Research and Development**. Boulder: Lynne Reiner Publishers, 1989. 237 pp.

Discusses the influence of the U.S. land-grant universities, agricultural experiment stations and Cooperative Extension Service on the agricultural institutions and policies and programs in developing nations. Some historical discussion of past policies and successes and failures regarding technological transfer.

1273. Conover, Milton. **The Office of Experiment Stations: Its History, Activities, and Organizations**. Baltimore: Johns Hopkins Press, 1924. 178 pp.

A dated but useful history. Includes some discussion of the National Alaskan and Insular Experiment Stations and notes the early activities of the experiment stations in Hawaii, Puerto Rico, Guam and the Virgin Islands.

1274. Cook, Melvin Thurston. **History of the First Quarter of a Century of the Agricultural Experiment Station at Rio Piedras, Puerto Rico**. Puerto Rico, University of Puerto Rico, Agricultural Experiment Station, 1937. 123 pp.

A history of the experiment station from 1910 to 1935. Includes a discussion of research on sugar cane, cotton, legumes, coffee, grasses, entomology, nutrition, plant breeding and livestock. In addition, this publication provides important indexes to the **Journal of the**

Department of Agriculture of Puerto Rico from volume 1 in January 1917 to volume 20 in October 1936 as well as to the station's circulars, bulletins and miscellaneous publications.

1275. Cooke, C.W., ed. **Agricultural Research, 1931-1981: A History of the Agricultural Research Council and a Review of Developments in Agricultural Science During the Last Fifty Years**. London: Agricultural Research Council, 1981. 367 pp.

Traces agricultural research in Great Britain, including the development of private agricultural research and the Rothamsted Experimental Station.

1276. Dabney, Charles W. Jr. "The Scientific Work of the Department of Agriculture." Office of Experiment Station, **Bulletin No. 24** (1895): 63-67.

Very brief description of the various USDA bureaus and divisions. Useful for perspective on the USDA's organization during the late nineteenth century.

1277. Danbom, David B. "The Agricultural Experiment Station and Professionalization: Scientists Goals for Agriculture." **Agricultural History** 60 (Spring 1986): 6-33.

During the late nineteenth and early twentieth centuries, scientists expected to improve agricultural productivity and the standard of living among farmers. In 1887, the Hatch Act created the agricultural experiment station system and enabled the application of science to agriculture.

1278. Eicher, C.L. "Building African Scientific Capacity for Agricultural Development." **Agricultural Economics: The Journal of the International Association of Agricultural Economics** 4 (June 1990): 117-43.

A critical examination of the influence of agricultural research centers south of the Sahara. Contends that direct technological transfer and donor-sponsored projects have failed to improve African agricultural scientific capabilities. Evaluates sixty years of African agricultural research and suggests changes for national and regional research institutions.

1279. Engel, P.G.H. "The Impact of Improved Institutional Coordination on Agricultural Performance: The Case of the Narino Highlands in Colombia." **Linkages Discussion Paper, International Service for National Agricultural Research No. 4**. The Hague, Netherlands: 1989. 23 pp.

Surveys the work of the Integrated Rural Development Programme in this region of the Andes from 1970 to the mid-1980s. Stresses the importance of technological transfer for improving agricultural productivity.

1280. Estey, Ralph H. "Publicly Sponsored Agricultural Research in Canada Since 1887." **Agricultural History** 62 (Spring 1988): 51-63.

A history of the federally funded agricultural experiment farms in Canada. Notes the origin of the Canadian experiment station system in 1906, including some discussion of the work in the provinces.

1281. Ferleger, Lou. "Uplifting American Agriculture: Experiment Station Scientists and the Office of Experiment Stations in the Early Years After the Hatch Act." **Agricultural History** 64 (Spring 1990): 5-23.

A general review of the early history of the U.S. experiment stations and the work of the USDA scientists. Contends the experiment station scientists substantially improved agricultural knowledge but failed to reach the farmer.

1282. Finlay, Mark R. "The German Agricultural Experiment Stations and the Beginnings of American Agricultural Research." **Agricultural History** 62 (Spring 1988): 41-50.

The first German experiment station was founded on September 28, 1858, but farmers and scientists could not agree whether research should be applied or pure. Eventually, agricultural scientists, particularly chemists, began to determine the research agenda, but practical research remained an important aspect of the German experiment station system, and it influenced the research agenda in the United States.

1283. Gates, Paul W. "The Morrill Act and Early Agricultural Science." **Michigan History** 46 (December 1962): 289-302.

Discusses the importance of the Morrill Act for establishing land-grant colleges and notes the curricula problems in the absence of systematic, scientific experimentation.

1284. Gayko, D. "Geschichte der Agrarwissenshaften der DDR." **Jarbuck fur Wirtschaftsgeschichte No. 4** (1987): 229-32.

Reports on a conference held in East Berlin on November 6-7, 1986, which emphasized the agricultural problems encountered in the German Democratic Republic as the nation moved from capitalism to socialism. The topics discussed include research in agricultural technology, agronomy, crop and livestock production and veterinary medicine. Emphasizes the period form 1945 to 1961.

1285. Gibb, J.A.C. **Agricultural Engineering Perspective: An Account of the First 50 Years of the Institution of Agricultural Engineers in Celebration of Its Golden Jubilee, 1938-1988**. Silsoe, Beds, United Kingdom: Instituion of Agricultural Engineers, 1988. 113 pp.

A survey of agricultural mechanization in relation to the role of the Institution of Agricultural Engineers in Great Britain.

1286. Gill, William R. **A History of the USDA National Tillage Machinery Laboratory: Agricultural Research Service, U.S. Department of Agriculture, Auburn, Alabama 36830, 1935-1985**. Auburn, AL: The Author, 1990. 255 pp.

Describes the work of the laboratory, particularly in relation to technological developments.

1287. Harding, T. Swann. **Two Blades of Grass: A History of Scientific Development in the United States Department of Agriculture**. Norman: University of Oklahoma Press, 1942. 352 pp.

A sympathetic history of the USDA based on departmental records, reminiscences of government scientists and personal observations. Dated but useful.

1288. Hightower, James. **Hard Tomatoes, Hard Times: A Report of the Agribusiness Accountability Project on the Failure of America's Land-Grant College Complex**. Cambridge: Schenckman, 1973. 268 pp.

A critical study of the experiment station system in the United States. Contends these institutions are devoted to improving the fortunes of large-scale, corporate farmers and the agribusiness industry while neglecting the needs of the small-scale, family farmer.

1289. Hoogen, J.P.M. van den, et al. "Institutional Development in Higher Agricultural Education." In **South-North Partnership in Strengthening Higher**

Education in Agriculture, edited by W. van den Bor, J.C.M. Shute and G.A.B. Moore, 23-41. Wageningen, Netherlands: Center for Agricultural Publishing and Documentation, 1989.

Discusses international cooperation for institutional development in agricultural education. Warns that direct transfer of Northern Hemisphere institutional approaches will not meet the educational needs south of the equator. Contends that institution building is a long-term affair.

1290. Humphries, F.S. "1890 Land-Grant Institutions: Their Struggle for Survival and Equality." **Agricultural History** 63 (Spring 1991): 3-11.

A general discussion of the black land-grant colleges in relation to curriculum, funding, legislation and racial discrimination.

1291. Kaimowitz, D. "Placing Agricultural Research and Technology Transfer in One Organization: Two Experiences from Colombia." **Linkages Discussion Paper, International Service for National Agricultural Research No. 3**. The Hague, Netherlands, 1989. 8 pp.

Discusses the experiences of the Colombian Agricultural Institute and the Colombian Coffee Growers Federation regarding agricultural research and technological transfer.

1292. Kerr, Norwood. **The Legacy: A Centennial History of the State Agricultural Experiment Stations, 1887-1987**. Missouri Agricultural Experiment Station, Special Report No. 350, 1987. 318 pp.

Discusses the development of agricultural research in a national context. An excellent administrative history that also notes the work of the individual stations. Emphasizes administrative relations between state and federal government, organizational structures, legislation and funding. Also notes changing public demands on agricultural research.

1293. Kimemia, J. "Agricultural Research Extension: Linkages with Small-scale Farmers: A Case Study in Kumba Corrido Agro-Ecological Zone, South West Province, Cameroon." **Farming Systems Analysis Paper No. 9**, 1990. 21 pp.

Discusses the influence of agricultural research on small-scale farmers and the ability of those farmers to gain institutional changes. Recommends that agricultural institutions support training workshops and encourage farmer participation.

1294. Kirkendall, Richard S. "The Agricultural Colleges: Between Tradition and Moderation." **Agricultural History** 60 (Spring 1986): 3-21.

The agricultural colleges encouraged capital-intensive farming based on the latest developments in science and technology. Rather than aid the family farm and maintain the agrarian tradition, these institutions supported the development of agribusiness.

1295. Lear, Linda J. "Bombshell in Beltsville: The USDA and the Challenge of 'Silent Spring.'" **Agricultural History** 66 (Spring 1992): 151-70.

In 1962, the USDA and the Agricultural Research Service were unprepared for the controversy created by Rachel Carson's attack on the pesticide industry. Agency scientists were committed to chemical solutions, and they were isolated from the public. They did not believe the public could understand complex scientific problems and shape the course of public policy.

1296. Marcus, Alan I. **Agricultural Science and the Quest for Legitimacy: Farmers, Agricultural Colleges, and Experiment Stations, 1870-1890**. Ames: Iowa State University Press, 1985. 269 pp.

An important study about the development of the experiment stations that preceded the Hatch Act in 1887. This book analyzes the long and often bitter struggle between farmers and scientists for the creation and control of agricultural experiment stations in the United States. Contends the research laboratory was not always central to agricultural experimentation, that agricultural chemistry and agricultural science are not synonymous and that the college-based agricultural experiment stations were founded as much on philosophy as on scientific and economic need.

1297. _____. "From State Chemistry to State Science: The Transformation of the Idea of the Agricultural Experiment Station, 1875-1887." In **The Agricultural Scientific Enterprise: A System in Transition**, edited by Lawrence Busch and William B. Lacy, 3-12. Boulder: Westview Press, 1986.

Discusses the difficulties encountered in establishing the state agricultural experiment stations as viable scientific institutions.

1298. _____. "The Ivory Silo: Farmer-Agricultural College Tensions in the 1870s and 1880s." **Agricultural History** 60 (Spring 1986): 22-36.

By the early 1870s, farmers considered the colleges to be primarily institutions for practical agricultural training, while the agricultural scientists at those colleges preferred to conduct research that could be applied to agricultural endeavors. Although the Education Act of 1890 attempted a compromise, the farmer-scientist conflict remained.

1299. Mayberry, B.D. **A Century of Agriculture in the 1890 Land-Grant Institutions and Tuskegee University, 1890-1990**. New York: Vantage Press, 1991. 271 pp.

A history of African American universities and colleges in the United States with special emphasis on agricultural education and Tuskegee University.

1300. Mathews, Eleanor. "Bibliographic Access to State Agricultural Experiment Station Publications." **Quarterly Bulletin of the International Association of Agricultural Librarians & Documentalists** 32 no. 4 (1987): 193-99.

A useful discussion of indexes and indexing services that provide access to the publications of the state agricultural experiment stations.

1301. Morrison, J.W. "Agricultural Achievements of the Prairie Experimental Farms, 1886-1986." **Prairie Forum** 11 (February 1986): 195-212.

Discusses the agricultural research conducted at the federal experimental farm stations in the Canadian Prairie Provinces since the Experimental Farm Station Act of 1886.

1302. Moore, Ernest G. **Agricultural Research Service**. New York: Frederick A. Praeger Publishers, 1967. 244 pp.

Authorized by the Secretary of Agriculture on November 2, 1953, the Agricultural Research Service consolidated most of the physical, biological, chemical and engineering research conducted by the USDA. This study provides a basic organizational history as well as discusses the relationship of the ARS to the public, Congress and other governmental agencies. Some discussion of the pesticide controversy created by Rachel Carson's **Silent Spring**.

1303. Nagarcenkar, R. "Dairy Research in India." In **Dairy Research: The Winds of Change. Proceedings, International Seminar, Glasgow, United Kingdom, 31 August 1 September 1989**, 75-87. Brussels: International Dairy Federation, 1989.

Describes the organization and research of the National Dairy Research Institute and the importance of the National Dairy Development Board. Notes the work of the Indian Council of Agricultural Research, particularly in relation to technological transfer.

1304. Powell, Fred W. **The Bureau of Plant Industry: Its History, Activities and Organization**. Baltimore: Johns Hopkins Press, 1927. 121 pp.

A dated, general history that includes a useful bibliography.

1305. Puls, Uta. "Vorgeschichte und Grundung der Deutschen Akademie der Landwirtschaftswissenschaften zu Berlin als Zentrum der Agrarforschung in der DDR (1945 bis 1951)." **Jarbuch fur Wirtschaftsgeschichte** [German Democratic Republic] no. 3 (1987): 49-68.

Discusses the transformation of the German Agricultural Society into the German Academy of Agricultural Sciences between 1945 and 1951.

1306. Pursell, Carroll W. Jr., "The Administration of Science in the Department of Agriculture, 1933-1940." **Agricultural History** 65 (Spring 1991): 168-72.

Between 1933 and 1940, the USDA enjoyed considerable achievement in agricultural research. Under the leadership of Henry A. Wallace, the agency built its research facility at Beltsville, Maryland, and began a program of basic research. Opportunism, however, rather than planning enabled much of this success.

1307. Rasmussen, Wayne D. "The 1890 Land-Grant Colleges and Universities: A Centennial Overview." **Agricultural History** 65 (Spring 1991): 168-72.

A brief review of the founding of the black land-grant colleges in the United States. Notes funding problems.

1308. Remenyi, J.V., and I.A. Coxhead. "A Survey of Farming Systems Research in Australia." **Journal of the Australian Institute of Agricultural Science** 52 no. 3 (1986): 135-43.

Describes the work of the Australian Centre for Agricultural Research for solving agricultural problems in developing nations.

1309. "Report on Workshop on Training of Rural Women in Post-harvest Loss Prevention, NIRD, Hyderbad, India, 8-14 August 1988." In **CIRDAP Training Series No. 39**, 1988. 89 pp.

Describes the work of the National Institute of Rural Development in developing training and educational instruction for women in relation to grain harvesting.

1310. Rodrigues, C.M. "A Pesquisa Agropecuaria Federal no Periodo Compreendido entre a Republica Velha e o Estado Novo." **Cadernos de Difusao de Technologia** [Brazil] 4 no. 2 (1987): 129-53.

Traces the development of agricultural research in Brazil from 1891 to 1945. Topics include sugar cane, coffee, fertilizers, pesticides, feeds, soils and plant diseases. Includes institutional relationships to scientific research.

1311. Rosenberg, Charles E. "Science and Social Values in Nineteenth Century America: A Case Study in the Growth of Scientific Institutions." In **Science and Values: Patterns of Tradition and Change**, edited by Arnold Thackray and Everett Mendelsohn, 21-42. New York: Humanities Press, 1974.

Discusses the role of social values in the creation of the agricultural experiment station system in the United States. Discusses the influence of European trained scientists, such as Evan Pugh and Samual W. Johnson.

1312. _____. "Science, Technology, and Economic Growth: The Case of the Agricultural Experiment Station Scientists, 1875-1914." **Agricultural History** 45 (January 1971): 1-20.

Describes the behavior of agricultural scientists and administrators within an institutional context. Emphasizes the period from the 1870s to 1914. Notes academic and research responsibilities. Provides examples from the agricultural experiment stations in Connecticut, Wisconsin and California.

1313. Rossiter, Margaret W. "The Organization of the Agricultural Sciences." In **The Organization of Knowledge in Modern America, 1860-1920**, edited by Alexandra Oleson and John Voss, 211-48. Baltimore: Johns Hopkins University Press, 1979.

Contends the farm interest groups, which believed science would improve productivity, gained annual appropriations from Congress to support agricultural research. The result was the creation of a vast bureaucracy of agricultural colleges and experiment stations across the nation.

1314. Sawyer, R.C. "Monopolizing the Insect Trade: Biological Control in the USDA, 1888-1951." **Agricultural History** 64 (Spring 1990): 271-85.

Criticizes the administrative structure of the USDA for hindering the biological control of artropod pests. Includes some discussion of activities in Hawaii.

1315. Schedvin C.B. **Shaping Science and Industry: A History of Australia's Council for Scientific and Industrial Research, 1926-49**. Sydney: Allen & Unwin, 1987. 374 pp.

A comprehensive institutional history that includes the discussion of agricultural research in Australia.

1316. Shah, J., et al. **Upgrading Technology**. Anand, India: Institute of Rural Management, 1990. 62 pp.

Discusses the work of the Institute of Public Administration and the Society for Appropriate Technology. Emphasizes decision making within an institutional setting in relation to the selection of agricultural technology for underdeveloped areas in India and South Asia.

1317. Sohne, W. "Abriss der Geschichte der Institute fur Landtechnik der Bundesrepublic." **Grundlagen der Landtechnik** 40 no. 1 (1990): 23-37.

Describes the history of the Agricultural Engineering Institute of the Technical University of Munich. Includes some discussion of other agricultural engineering institutes in the Federal Republic of Germany.

1318. Smith, Maryanna S. **A List of References for the History of the United States Department of Agriculture**. Davis, CA: Agricultural History Center, University of California, Davis, 1974. 88 pp.

An excellent place to begin research on the USDA. This source, however, should be supplemented with the recent indexes for **Agricultural History** and **America: History and Life**.

1319. Sollig, Valentin, and Elke Melichar. "Die Entwicklung der Lehre auf dem Gebiete der Agrarwissenschaften unter Besonderer Berucksichtigung der Tierproducktion and der Wilhelm-Pieck-Universitat Rostock Von 1946 bis sur Gegenwart." **Gesellschaftswissenschaftliche Reihe** [Democratic Republic of Germany] 35 no. 1 (1986): 45-53.

    Discusses the development of teaching agricultural science at the University of Rostock (East Germany) since World War II. Emphasizes the program for livestock production.

1320. Stevenson, John A. "Plants, Problems, and Personalities: The Genesis of the Bureau of Plant Industry." **Agricultural History** 28 (October 1954): 155-62.

    A survey of the first fifty years of the Bureau from its founding on July 1, 1901. Good for material on the major individuals and their activities in the Bureau.

1321. Stewart, Robert. **Seven Decades that Changed America: A History of the American Society of Agricultural Engineers, 1907-1977.** St. Joseph, MI: American Society of Agricultural Engineers, 1979. 440 pp.

    An institutional history that emphasizes organization rather than scientific research.

1322. Storer, Norman. "Science and Scientists in an Agricultural Research Organization: A Sociological Study." Ph.D. diss., Cornell University, 1961. 237 pp.

Discusses the difference between applied and basic research at the state agricultural experiment stations in terms of sociology. Contends that basic research scientists are younger and come from higher socio-economic backgrounds and perceive their work and profession differently than those scientists engaged in applied research.

1323. Street, Donald R. "Spanish Antecedents to the Hatch Act Experiment Station System and Land Grant Education." **Agricultural History** 62 (Spring 1988): 27-40.

Discusses the similarities in research and extension between the land-grant college system in the United States and Spanish economic societies that organized during the eighteenth century.

1324. "Strengthening Bolivia's Agricultural Research and Technology Transfer System." **ISNAR Report No. R45s**, 1989. 83 pp.

Studies the organization of the Research Center for Tropical Agriculture and the Bolivian Institute of Agricultural Technology. Suggests reforms necessary to improve research and technological transfer.

1325. Trigo, E.J. "Agricultural Research Organization in the Developing World: Diversity and Evolution." In **Policy for**

Agricultural Research, edited by Vernon W. Ruttan and Carl E. Pray, 251-81. Boulder: Westview Press, 1986.

Reviews the history of public and private agricultural research organizations in Asia, Latin America and Africa.

1326. True, A.C. "Agricultural Experiment Stations in the United States." In **Yearbook of Agriculture, 1899**, 513-48. Washington, D.C.: United States Department of Agriculture, 1899.

A general history of the origin and development of the experiment station system. Written for the general reader, but a useful introduction.

1327. _____. "A History of Agricultural Experimentation and Research in the United States, 1607-1925. Including a History of the United States Department of Agriculture." United States Department of Agriculture, **Miscellaneous Publication No. 251**, 1937. 321 pp.

An encyclopedic catalog of the organizations and institutions that shaped late nineteenth-century agricultural science. Includes sketches of the legislative attempts to pass a national experiment station bill. A good, factual reference.

1328. _____, and D.J. Crosby. "Agricultural Experiment Stations in Foreign Countries." United States Department of Agriculture, Office of Experiment Stations, **Bulletin 112**, 1902. 276 pp.

Lists 798 agricultural experiment stations and similar institutions. Arranged alphabetically by country and city. Provides a brief description of these institutions located in 51 nations. Notes the research, staffing and budgets of each.

1329. Waggoner, Paul E. "Research and Education in American Agriculture." **Agricultural History** 50 (January 1976): 230-47.

Since the late nineteenth century, the agricultural experiment stations have had disputes with the USDA over their independence and with the agricultural colleges over the responsibilities of their scientists.

1330. Weber, Gustavus Adolphus. **The Bureau of Chemistry and Soils: Its History, Activities, and Organization**. Baltimore: Johns Hopkins Press, 1928. 218 pp.

An institutional history that discusses the major research activities of the chemistry and soils units as well as the organization of the bureau.

1331. _____. **The Food, Drug and Insecticide Administration: Its History, Activities, and Organization**. Baltimore: Johns Hopkins Press, 1928. 134 pp.

A dated but useful survey that discusses early research activities as well as food and drug regulations. This source should be supplemented with the work of Harvey Young.

1332. White, G.A., H.L. Shands, and G.R. Lovell. "History and Operation of the National Germplasm System." **Plant Breeding Reviews** 7 (1989): 5-56.

Includes a brief description of the collections for small grains, soybeans, cotton and potatoes as well as quarantine regulations and procedures.

1333. Wiest, Edward. **Agricultural Organization in the United States**. Lexington: University of Kentucky, 1923. 618 pp.

Discusses the organization and responsibilities of the USDA and notes various private agricultural groups such as the Farm Bureau and breeders associations. Economic in approach.

1334. Williams, Thomas T., and Handy Williamson, Jr. "Teaching, Research, and Extension Programs at Historically Black (1890) Land-Grant Institutions." **Agricultural History** 62 (Spring 1988): 244-57.

In 1890, Congress passed the second Morrill Act to support agricultural education and research at black land-grant colleges, but insufficient funding and racial prejudice continued to retard the development of these schools. By the late twentieth century, however, 17 black land-grant colleges and universities, including Tuskegee University, provided high-quality undergraduate and graduate programs in agricultural research.

1335. Wilson, M.L. "Survey of Scientific Agriculture." **Proceedings of the American Philosophical Society** 86 (1942): 52-62.

Brief descriptions of various agricultural issues in the United States. Notes the publications of the American Philosophical Society between 1771 and 1809 that concerned agriculture.

1336. Wiser, Vivian, and Wayne D. Rasmussen. "Background for Plenty: A National Center for Agricultural Research." **Maryland Historical Magazine** 61 (December 1966): 283-304.

Traces the origin and development of he USDA research center at Beltsville, Maryland.

1337. Woods, A.F. "The Development of Agricultural Research and Education Under the Federal Government." **Scientific Monthly** 36 (1933): 15-34.

Traces the work and accomplishments of the USDA from 1862 to 1930.

CHAPTER XIX

AGRICULTURAL POLICY

1338. Antle, J.M., and S.M. Capalbo. "Pesticides and Public Policy: A Program for Research and Policy Analysis." In **Agriculture and the Environment**, edited by Tim T. Phipps, Pierre R. Crosson and Kent A. Price, 155-74. Washington, D.C.: National Center for Food and Agricultural Policy, Resources for the Future, 1986.

    A general discussion of pesticide regulations and environmental pollution in American agriculture. Notes health hazards. A useful introduction.

1339. Bayma, B.A., Jr. "Technology Transfer: A Public Policy Issue." **Journal of Technology Transfer** 3 (Spring 1979):43-51.

    Provides an historical overview of American technological exports. Argues that technology transfer is a public policy issue in the United States.

1340. Beck. Robert E. "Pesticides and the Law." **North Dakota Quarterly** 31 no. 1 (1969): 49-64.

Discusses the dangers of pesticides and analyzes statutory regulations. Contends enforcement is inadequate and that the common law could be utilized to provide greater public protection,

1341. Black, K.S., and F.M. Doepel. "Legislation Policy Attitudes: Impacting ARF Management." **Lab Animal** 14 (April 1985): 40-43.

A brief review of animal welfare legislation regarding laboratory work. Emphasizes regulations in Great Britain

1342. Bogue, Allan G. "Comment: Policy Making in the SPA." **Agricultural History** 64 (Spring 1990): 244-51.

Suggests areas for further research. Merits consideration by anyone working in this area.

1343. Bonnen, J.T. "A Century of Science in Agriculture: Lessons for Science Policy." In **Policy for Agricultural Research**, edited by Vernon W. Ruttan and Carl E. Pray, 105-37. Boulder: Westview Press, 1987.

Reviews American agricultural policy regarding pure and applied research. Evaluates the benefits and problems of support from the public and private sectors.

1344. Bosso, Christopher J., Norman J. Vig, and Duane Windsor. **Pesticides and Politics: The Life Cycle of a Public Issue**. Pittsburgh: University of Pittsburgh Press, 1987. 294 pp.

Studies policy making since 1947. An excellent introduction to the politics of pesticide regulation.

1345. Bul'maga, L.E. "Nauchno-Teknicheskii Progressa v Sel'Skom Khoziaistve Moldavskoi SSR." **Voprosy Istorii** (Soviet Union) no. 1 (1975): 23-31.

The author reviews the scientific and technological policy and progress of Soviet agriculture from 1959 until 1970. The geographic focus is the Moldavian Soviet Socialist Republic.

1346. Busch, Lawrence, and William B. Lacy. **Science, Agriculture, and the Politics of Research**. Boulder: Westview Press, 1983. 303 pp.

Examines the factors that influence the choice of research problems by scientists. Emphasizes the process and context in which public agricultural research is conducted in the United States. Written for the general audience, but scholars will find this study useful.

1347. Buttel, Frederick H., and Lawrence Busch. "The Public Agricultural Research System at the Crossroads." **Agricultural History** 62 (Spring 1988): 303-24.

A provocative review of the problems faced by the agricultural experiment stations and land-grant colleges in relation to research agendas, funding, extension and curricula. Suggests changes necessary to meet contemporary problems and to put the traditional and no longer viable methods of operation behind as the nation faces the twenty-first century.

1348. Charrym, A., and J.L. Dillon. "Structuring National Research with a Farming Systems Perspective for the Tropical Savannas of Colombia. **Quarterly Journal of International Agriculture** 28 no. 3/4 (1989): 315-25.

Argues that research to improve livestock raising must be based on a policy that recognizes the importance of environmental and organizational as well as technical issues. Researchers must consider the limited resource base, sustainability and the beliefs of farmers, if they are to provide relevant solutions for agricultural problems.

1349. Crocker, Davis P. "The Pesticide Controversy--A Citizen's Perspective." **Proceedings of the Western Society of Weed Science** 37 (1984): 85-93.

Discusses regulatory legislation and registration procedures for pesticides, particularly in relation to the public concern about the safety of agrochemicals.

1350. Datta, Rakhal. "Technology Choice in Collectivized Agriculture: Farm Mechanization Policy of the People's Republic of China." **China Report** [India) 16 no. 5 (1980): 3-30.

A discussion of Chinese agriculture policy since the 1950s. The policy of "agriculture first," which emphasized multiple cropping was established in 1962. Since that time, governmental policy has forced Chinese farmers to mechanize.

1351. Evenson, R.E., and G. Ranis, ed. **Science and Technology: Lessons for Development Policy**. Boulder: Westview Press, 1990. 391 pp.

The success of technological transfer depends on the informed adoption of specific technologies and policies for local conditions. Agriculture provides several examples, particularly for East Asia.

1352. Fisher, S.W. "The Roots of Controversy: A Historical View of Pesticide Regulation." **American Nurseryman** 165 (January 1, 1981): 89-90, 92-94, 96-99.

Traces pesticide regulation in the United States to the early twentieth century. Emphasizes the importance of the Pesticide Act of 1910, which protected farmers against fraud by ensuring product purity. This legislation, however, did not address matters of environmental contamination.

1353. Gabrielian, Egnara G. "Greorgi Dimitrov i Razvitieto na Sialskostopanskata Nauka v Bulgariia." **Istoricheski Pregled** [Bulgaria] 38 no. 2 (1982): 77-83.

After World War II, the application of science to agriculture in Bulgaria received support from both the government and the Soviet Union. Georgi Dimitrov, leader of the Bulgarian Communist Party, played an instrumental role in shaping governmental agricultural policy.

1354. Gendel, Steven M., ed. **Agricultural Bioethics: Implications of Agricultural Biotechnology**. Ames: Iowa State University Press, 1990. 357 pp.

A collection of symposium papers that address the moral and ethical implications of biotechnology. A provocative survey of new considerations for agricultural policy makers.

1355. Guerin, Charles Leo. "The Federal Regulation of Pesticides." Ph.D. diss., University of Massachusetts, 1983. 484 pp.

Contends that pesticide regulation in the United States is inadequate because congressional agricultural committees have favored weak laws and because the manufacturers and pesticide user groups support lenient legislation. Argues that the land-grant colleges emphasize research for chemical control rather than the development of environmentally sound technologies.

1356. Heichel, G.H. "Communicating the Agricultural Research Agenda: Implications for Policy." **Journal of Production Agriculture** 3 no. 1 (1990): 20-24.

Discusses the criticism that began in 1973 of public agricultural research in the United States. Emphasizes four areas of criticism: 1) absence of a consensus and agenda among agricultural scientists; 2) an inadequate relationship between pure and applied research; 3) poor communications between agricultural scientists and policy makers; and 4) the system for rewards.

1357. Huang, H.T. The USDA's Competitive Grants Program and Agricultural Research." **Agricultural History** 62 (Spring 1988): 270-95.

Analyzes the effects of the work of the Competitive Research Grants Office in 1977 to make agricultural research more applicable to contemporary problems. Because USDA and agricultural experiment station scientists must compete with other researchers from different institutions for these funds, this policy has caused great criticism by the public scientists. Nevertheless, this policy has improved basic research in the agricultural sciences, especially the plant sciences, and it has increased the number of scientists working on agricultural projects.

1358. Kerr, Norwood Allen. "Institutionalizing the New Agenda: The State Agricultural Experiment Stations, 1977-1981."

Agricultural History 62 (Spring 1988): 279-95.

Summarizes the content of Title XIV of the Food and Agriculture Act of 1977. This portion of the legislation not only posed new areas for research but based funding on a competitive basis and removed it from the total domination of the USDA and agricultural experiment stations. Although Title XIV has been modified, the state agricultural experiment stations have not regained total control of funding for agricultural research.

1359. Kinteh, S. "A Review of Agricultural Policy Before and After Adjustment." **Working Papers on Commercialization of Agriculture and Nutrition**, No. 4, 5-26. Washington, D.C.: International Food Policy Research Institute, 1990.

Describes Gambia's agricultural policy since the mid-1970s. Evaluates the role of public institutions and the problems and possibilities that agricultural policy makers confront, especially in relation to mechanization.

1360. Kirichenko, N.K. "Partiinoe Rukovodstvo Uskoreniem Nauchno Tekhnicheskogo Progressa v Sel'Skom Khoziaistve." **Voprosy Istorii** [Soviet Union] no. 3 (1974): 34-46.

During the early 1970s, the Soviet government attempted to increase production by providing greater support for agricultural industry, education and research. This policy emerged from the 24th Congress.

1361. Koehn, P.H. "Agricultural Policy: Bureaucratic interests, Environmental Impact, and Socio-Economic Outcomes." In **Public Policy and Administration in Africa: Lessons from Nigeria**, 85-108. Boulder: Westview Press, 1990.

Discusses the failures of the Bakalori irrigation and Funtua agricultural development projects in northern Nigeria. Contends that agricultural policy, influenced by foreign governments and multinational corporations, did not address the basic needs of small-scale farmers.

1362. MacIntyre, A.A. "Why Pesticides Received Extensive Use in America: A Political Economy of Agricultural Pest Management to 1970." **Natural Resources Journal** 27 (Summer 1987): 533-78.

Discusses the political and institutional networks that supported the use of pesticides in the United States. Views pesticide development, use and regulation in conspiratorial terms. Suggests the federal government should be cautious in permitting the use of pesticides because the safety of each is not easily proven, nor is the research data necessary for informed policy making always available.

1363. Malish, Anton F. "Soviet Agricultural Policy in the 1980s." **Policy Studies Review** 4 no. 2 (1984): 301-10.

In January 1980, the United States placed an embargo on agricultural commodities destined for the Soviet economy. The Soviet Union responded by attempting to gain self-sufficiency in meat production, and it achieved some success. Thereafter, American grain exports to the USSR decreased but U.S. exports of agricultural technology increased.

1364. March, B.E. "Bioethical Problems: Animal Welfare, Animal Rights." **BioScience** 34 (November 1984): 613-20.

Provides historical background about public attitudes toward animals. An important study in a time when scientists, agriculturists and nonvegetarians find their activities challenged. Discusses animal rights and legislation.

1365. Marcus, Alan I. "The Wisdom of the Body Politic: The Changing Nature of Publicly Sponsored American Agricultural Research Since the 1830s." **Agricultural History** 62 (Spring 1988): 4-26.

An excellent overview of publicly sponsored research in relation to the state agricultural experiment stations and the USDA. Notes the loss of power and prestige by these institutions with the rise of the National Science Foundation and the National Institutes of Health as research sponsoring agencies for work in the life sciences after 1930.

1366. Musella, Luigi. "La Modernizzazione Tecnica del Mezzogiorno Rural e L'Azione del Ministero di Agricoltura, 1878-1896." **Studi Storici** [Italy] 29 no. 1 (1988): 207-30.

A study of the technical modernization of rural southern Italy during the late nineteenth century. Although the Ministry of Agriculture established a supportive agricultural policy in 1878, the Italian government faced many difficulties convincing landowners and peasants to adopt new agricultural technology.

1367. O'Brien, J.W. "Science Policy, Biotechnology and American State Government: Recommendations for State Action." **Studies in Technology and Social Change, No. 12**. Ames: Technology and Social Change Program, Iowa State University, 1989. 16 pp.

Discusses state supported science policy in the United States by using agricultural biotechnology as an example. Analyzes state science policy since World War II, and describes the political context of agricultural biotechnology in relation to public and private constituencies.

1368. Offut, S., et al. "Technology Policy and Agriculture." **American Journal of Agricultural Economics** 73 no. 3 (1991): 876-904.

A collection of seven symposium papers that discuss the relationships between public institutions and private industry, plant variety and patent protection and international technological transfer. Includes references to the United Kingdom, France and Germany.

1369. Perkins, J.H. **Insects, Experts, and the Insecticides Crisis: The Guest for New Pest Management Strategies**. New York: Plenum Publishing Corp., 1982. 304 pp.

Discusses the cultural forces that affect the creation of scientific and technical knowledge. Emphasizes the debate about the safety and regulation of insect controls from 1943 to 1980.

1370. Pijawka, David K. "A Comparative Study of the Regulation of Pesticide Hazards in Canada and the United States." Ph.D. diss., Clark University, 1983. 263 pp.

Examines the major regulatory legislation for pesticides in each country. Emphasizes four case studies. A good place to begin research, especially in relation to Canada.

1371. Poats, S.V. "Farming Systems Research in Some Other West African Countries." **IDRC Manuscript Report, No 172**, 88-93. Ottawa, Canada: International Development Research Centre, 1989.

Surveys contemporary agricultural research and policy in the Ivory Coast, Sierra Leone, Burkina Faso, Benin, Togo, Cameroon, Mauritania, Niger and Ghana.

1372. Rasmussen, Wayne D. "The Influences of Policy on Agricultural Research and Technology: An Historical Perspective." **Agricultural Science Policy in Transition**, edited by V. James Rhodes, 183-201. Bethesda, MD: Agricultural Research Institute, 1986.

Reviews the history of research policy in the United States Department of Agriculture. A good overview of historical developments.

1373. Rosenberg, Charles E. "Rationalization and Reality in the Shaping of American Agricultural Research, 1875-1914." **Social Studies of Science** 7 (November 1977): 401-22.

Contends that pure and applied agricultural scientists compromised their work to meet the needs of agricultural clients during the nineteenth and twentieth centuries. Generous federal funding for each research area provided for the needs of a variety of agricultural scientists and prevented animosity between each group.

1374. Rossiter, Margaret W. "Science and Public Policy Since World War II." **Osiris** [Belgium] 1 (1985): 273-94.

Calls for greater study of the interrelationships between science and public policy. Surveys studies in eight areas and suggests topics for research. Historians of agricultural science and technology will find her discussion of environmental issues suggestive for further work.

1375. Singer, Peter. "Ethics and the New Animal Liberation Movement." In **In Defense of Animals**, edited by Peter Singer, 1-10. New York: Basil Blackwell, 1985.

Describes past thought concerning the treatment of animals, including that of the Greeks, Hindus, Buddhists and Christians. Contends the treatment of animals and humans should be equal if the situation justifies it. Focuses on the animal liberation movement of the 1970s.

1376. Swazey, J.P. "Protecting the 'Animal of Necessity': Limits to Inquiry in Clinical Investigation." In **Limits of Scientific Inquiry**, edited by Gerald Holton and Robert S. Morison, 129-45. Mew York: W.W. Norton, 1979.

A brief history of the ethics, laws and research policies that apply to laboratory investigations with animals. This study will be suggestive for further research in agricultural and veterinary science.

1377. Vatn, A. "Teknologisk Utvikling og Insatitusjonel Endring: Ein Studie av Sider ved Norsk Jordbrukspolitikk, 1920-1980." Ph.D. diss., University of Norway, 1983. 418 pp.

A study of technological development and institutional change in relation to the formulation of agricultural policy in Norway from 1920 to 1980.

1378. Vyvyan, John. **In Pity and In Anger: A Study of the Use of Animals in Science**. Marblehead, MA: Micha, 1988. 161 pp.

Essentially a historical study of vivisection during the nineteenth century.

1379. Wiser, Vivian. "Public Policy and USDA Science, 1879-1913." **Agricultural History** 64 (Spring 1990): 24-30.

Reviews the administration of James Wilson as Secretary of Agriculture in relation to improvements in livestock and plant research. Wilson strongly supported the agricultural experiment stations and continuing education for scientists. He substantially strengthened the importance of science in the USDA.

1380. Zebich, Michele. "The Politics of Nutrition: Issue Definition, Agenda-Setting and Policy Formulation in the United States." Ph.D. diss., University of New Mexico, 1979. 274 pp.

Examines the factors that influence nutrition policy. Emphasizes policy orientation from 1974, with the creation of the National Nutrition Plan by the Senate Select Committee on Nutrition and Human Needs, to passage of the Food and Agriculture Act of 1917.

INDEX

[All Numbers Refer to Citations]